LEVITATING TRAINS AND KAMIKAZE GENES

LEVITATING TRAINS
and
KAMIKAZE GENES
*Technological Literacy
for the Future*

RICHARD P. BRENNAN

WILEY

Wiley Popular Science

JOHN WILEY & SONS, INC.

New York   Chichester   Brisbane   Toronto   Singapore

For
Carolyn F. Brennan, with love

The Library of Congress has cataloged the hardcover thus:

Brennan, Richard P.
    Levitating trains and kamikaze genes : technological literacy for
the 1990s / Richard P. Brennan.
      p.   cm. — (Wiley science editions)
    Includes bibliographical references.
    ISBN 0-471-62295-8
    1. Technology. I. Title. II. Series.
T47.B733   1990
600—dc20   89-37349

    ISBN 0-471-07902-2 (paper)

Printed in the United States of America

10 9 8 7 6 5 4 3 2

CONTENTS

PREFACE

IF YOU BELIEVE IN BIG FOOT, the Loch Ness Monster, UFOs, aliens, or the lost city of Atlantis, you have picked up the wrong book. You may even be in the wrong bookstore. You are not alone, mind you. More than one-third of college students polled in Texas, Connecticut, and California in 1988 by University of Texas researchers said they believed in these things. Overall, some 37 percent of these students said they believed in ghosts, and 39 percent said it is possible to communicate with the dead.

What really surprised the researchers were the number of students who believed what the researchers called "cult beliefs" or unproven pseudoscientific theories. Francis Harrold, a professor of archeology at the University of Texas who helped create and conduct this survey, said, "We all agree that for a leading scientific nation, this is not a good sign of the effectiveness of our scientific education."

Another 1988 survey of 2100 students on 41 campuses around the country conducted for the National Science Foundation revealed that a large percentage of students believed that in the early history of Earth, cave dwellers battled marauding dinosaurs. (Dinosaurs roamed the Earth for over 200 million years, but they disappeared about 60 million years before humans appeared.) Large percentages of these same students believe that fossils are the remains of animals that didn't make it onto Noah's ark.

The college students did better on the survey than the general public, at least as measured by its responses to similar questions asked in a Gallup poll taken in 1982. But the student survey did reveal that education is no guarantee of science literacy.

The bad news for these eager students is that they are technological illiterates. The world is not as they think it is. This book is about the way the world is; a world governed by the immutable laws of physics (whether we fully understand them or not); a macrocosm of almost unimaginable reaches containing billions and billions of galaxies each containing billions and billions of stars; and a subatomic microcosm of equally unimaginable smallness containing subnuclear particles, most of which live for only a billionth of a billionth of a second. In short, this book is about reality, about science and about technology.

If that doesn't scare you off, nothing will.

Levitating trains, kamikaze genes, space travel, artificial intelligence, new power sources, life-enhancing medical advancements, life-threatening ozone holes, star wars weapons systems—this book is about all of these technologies and more.

Face it—most of us who are nonscientists or nonengineers, whatever our educational credentials, have large gaps in our knowledge of science and technology. There are important areas about which we know nothing; even if we completed a science requirement in school, we are still a bit hazy on the latest developments in that particular subject and certainly ignorant about other areas of science and technology.

This book is designed to fill in the gaps and give you a solid background in the world of science and technology. In short, it's intended to provide the technical knowledge we all must have to succeed in a high-tech world.

This book is also intended to help you avoid the traps of technical illiteracy. It is ironic that we live in an age of great technical and scientific advancements, while at the same time scientific and technical ignorance prevails.

A 1984 Gallup poll found that 55 percent of teenagers believe that astrology works. We live in an age of channelers talking to long-dead ancestors, UFO abductions about every other week, and gurus spouting all kinds of New Age nonsense. In sum, today's world is a high-tech world *and* at the same time a regression to the most primitive sort of ignorant superstitions.

It is as if technological illiteracy was a black hole that sucks into itself all the pseudosciences, all the mumbo jumbo and nonsense floating about out there.

The cutting edge of science moves very rapidly these days, and the result for many of us is often technological shock. The intelligent response to this shock is not to avoid technology but rather to identify those areas where some catch-up effort is in order.

What then is technological literacy? What basic engineering or scientific concepts should be understood by all nonscientists or nonengineers? This book sets out to define what such literacy encompasses and to explain basic concepts in space exploration, biotechnology, computer science, energy technology, and superconductivity. The book also explores innovations and dangers in medical, environmental, transportation, and defense technologies.

In each case, not only do I discuss a new and important technology but also I describe the basic principles of science involved. In this way you can use technology to help understand our rapidly changing world. For instance, Einstein's concepts of space, time, and gravity are introduced as part of the discussion of space exploration. The basic principles of physics appear in the examination of lasers, solar energy, nuclear power, and superconductors. The First and Second Laws of Thermodynamics are integral to any examination of advanced engines or new power sources for supercars of the future. Some basic biology is necessary to comprehend the implications of the current biotechnical revolution.

This is admittedly an ambitious undertaking, but note that this book includes only what an otherwise educated person should know about these technologies and no more. It's a condensation of mountains of research and expert knowledge.

In the course of my long career in science writing I always found it exciting that I, a nonscientist and nonengineer, could comprehend and appreciate the concepts my enthusiastic technical colleagues patiently explained to their editor. This book is in part an effort to share that sense of discovery.

One of the challenges of good science writing is striking the right balance between the general and the detailed. Scientists and engineers often write books that are too technical by far for general readers. Journalists, on the other hand, often write material that is too superficial. Balance is the goal here.

I offer in this book a purposely lean but extraordinarily broad *tour d'horizon* of the technological world—again, what you should know to be technologically literate.

I am grateful to Carolyn F. Brennan for the illustrations that accompany the text as well as for her critical reviews of the manuscript. I am also indebted to my editor, David Sobel, for his many contributions.

—Richard Brennan

INTRODUCTION

RECENTLY THE DEBATE over "cultural literacy" in our society grabbed many headlines across the nation. What, if any, core of basic knowledge should be an essential part of everyone's education? It is an important issue and it is still being debated.

But what about "technical literacy"? Are there not basic technical concepts about our world and our society that every citizen should know, particularly when many of the important questions that confront us today are essentially technical in nature; vital issues concerning our environment, our health, our transportation, our communications, and our national priorities?

The English writer/scientist C.P. Snow touched on this matter in his famous essay *Two Cultures* some years ago. His views on the gulf that exists between the scientist and nonscientist and how fatal the lack of communications could be became a much-discussed issue of the time.

Snow, who spent his days as a working scientist and his evenings with his literary friends, became acutely aware of the dichotomy and lack of understanding between the two cultures (science and liberal arts). He pointed out that his scientist/engineer friends, while highly intelligent people, were woefully ignorant of what we have come to call the "liberal arts." It is as if, Snow said, the whole literature of traditional culture didn't seem to be relevant to their interest. As a result, their imaginative understanding is less than it could be. They are self-impoverished.

But this was not a one-sided story. Here is how Snow addressed the opposite ignorance.

> But what about the other side? They are impoverished too—perhaps more seriously, because they are vainer about it. They still like to pretend that the traditional cultural is the whole of "culture," as though the natural order didn't exist. As though the exploration of the natural order was of no interest either in its own value or its consequences.
>
> As with the tone-deaf, they don't know what they miss. They give a pitying chuckle at the news of scientists who have never read a major work of English literature. They dismiss them as ignorant specialists. Yet their own ignorance and their own specialisation is just as startling. A good many times I have been present at gatherings of people who, by the standards of traditional culture, are thought highly educated and who have with considerable gusto been expressing their incredulity at the illiteracy of scientists. Once or twice I have been provoked and have asked the company how many of them could describe the Second Law of Thermodynamics. The response was cold; it was also negative. Yet I was asking something which was the scientific equivalent of: Have you read a work of Shakespeare's?
>
> I now believe if I had asked an even simpler question—such as, What do you mean by mass, or acceleration, which is the equivalent of saying, Can you read?—not more than one in ten of the highly educated would have felt that I was speaking the same language.*

Snow obviously thought that technical illiteracy was as deplorable as any other kind. "The worst of crimes," he wrote, "is innocence."

Innocence prevails in our educational system at the moment. The scores of American students ranging in all grades of elementary through high school on international science tests reflect poorly on science education in the United States. The 1988 Second International Science· Study conducted by the National Science Foundation showed U.S. students placed in the middle at best and more often at the bottom among test takers around the world.

---

*From *Two Cultures: And A Second Look*, C. P. Snow. New York, Cambridge University Press, 1959. Reprinted by permission.

The NSF study concluded that for a technologically advanced country, it would appear that a reexamination of how science is presented and studied in the United States is required.

Why is technological literacy important? Because we all suffer if ignorance prevails. To make matters worse, fewer and fewer of our young people seem to be interested in science as a career. According to Eric Bloch, the former director of the National Science Foundation, in a 1988 speech to the American Physical Society, the number of students enrolled in science and technical courses has diminished in each of the last few years.

An additional irony, Bloch has pointed out, is that the participation of Americans in science and engineering has been declining in the face of growing demand. The demand has been met in large part by foreign students, who now earn more than one-fifth of the science doctorates, one-third of the mathematics doctorates, and more than half of the engineering doctorates awarded each year by U.S. universities.

U.S. overseas competitors—Japan, West Germany, France, and the United Kingdom—have increased the proportion of their labor forces in research and development, although we have not. In 1965, for example, each of these four countries had only about one-third as many scientists and engineers in proportion to their population as did the United States. By 1984, Japan had nearly matched the United States' level of professionals, and the others had grown to roughly two-thirds the U.S. level.

In his novel *Player Piano*, Kurt Vonnegut envisions a world divided into a knowledgeable technical elite that controls and rules, a core of robots that perform the work, and a great mass of ignorant unemployed who live on a sort of dole and are kept amused by games, sports, and lotteries. Although Vonnegut was writing futuristic fiction, it may be about time we all pulled up our collective technological socks.

TEST YOUR TECHNOLOGICAL LITERACY

How do you measure up? This is your chance to evaluate your own technological literacy in private and without embarrassment. Nobody will know how well you perform on this test except you, and you alone can decide what to do about any fuzzy areas.

This test contains 50 questions covering the subjects emphasized in this book: space, biotechnology, computer technology, environmental issues, energy, superconductivity, high-tech medicine, transportation technology, and weapons and arms control technology. Each question is multiple choice, and four possible answers are listed for each question.

When you have completed the test, turn to the Appendix for the answers. They are intended to provide enough information for you to determine why your response was correct or incorrect.

You may, if you wish, grade yourself in accordance with the following guideline:

Scoring Guideline

Score two points for each correct answer.

90–100   *Excellent.* Congratulations, you're certified technologically literate.

80–88   *Good.* You're well on your way to technological literacy.

70–78   *Fair.* Not too bad, but there are gaps in your technical knowledge.

60–68   *Poor.* There is room for improvement in some areas of your technical knowledge.

Below 60   *It is time for you to pull up your technological socks.* A reading program and/or a subscription to a science/technology magazine is in order.

After you've read the book, come back and take the test again, and see how technologically literate you have become.

Space

1. A light year is
   a. the accepted measure of time in outer space.
   b. the distance between the Sun and Earth.
   c. the distance light travels in one year.
   d. the time it takes for light from the Sun to reach Earth.

2. What is a star?
   a. Any object in the sky that is discernible by telescope
   b. A self-luminous gaseous body in space that generates energy by means of nuclear fusion at its core
   c. A moon or a planet
   d. A heavenly body, especially a planet, having influence on one's fortune and destiny

3. Which of the following is the correct order of the planets in our solar system in terms of average distance from the Sun?
   a. Mercury, Venus, Earth, Mars, Jupiter, Saturn, Uranus, Neptune, Pluto
   b. Earth, Venus, Mars, Mercury, Jupiter, Saturn, Uranus, Neptune, Pluto
   c. Mercury, Venus, Earth, Jupiter, Saturn, Uranus, Neptune, Pluto, Mars
   d. Mars, Mercury, Venus, Earth, Pluto, Saturn, Uranus, Jupiter, Neptune

4. The Big Bang theory is associated with
   a. violent geological upheavals such as volcanic eruptions or earthquakes.
   b. explosive volatility of certain chemical compounds.
   c. the cataclysmic birth of the universe.
   d. vents in the Earth's crust through which molten rock, ashes, and lava are sometimes ejected.

5. A black hole can be described as
   a. the vast empty area between galaxies in the universe.
   b. the black empty space between stars.
   c. the dense matter left over after the collapse of a star.
   d. a new star radiating only a small amount of light.

Biotechnology

1. DNA (deoxyribonucleic acid) is
   a. the main component of chromosomes and the material that transfers genetic characteristics in all life forms.
   b. found only in complex, higher order life forms including *Homo sapiens.*
   c. the chemical substance found only in plants.
   d. found only in mammals.

2. What are chromosomes?
   a. One-celled organisms that are the most primitive form of life
   b. Collections of genes
   c. A chromium alloy
   d. Multicolored algae

3. Genetic engineering is
   a. the use of biological techniques to rearrange genes: to remove, add, or transform them from one organism or location to another.
   b. the attempt to synthesize life from nonliving materials.
   c. the attempt to improve the human population by discouraging reproduction by individuals with genetic defects.
   d. the process by which sex cells are formed.

4. Which of the following statements is true?
   a. The government has given scientists permission to transfer foreign genes into humans.
   b. Inserting foreign genes into humans is banned in the United States.
   c. There are scientific and technical reasons preventing foreign genes from being transfered into human beings.
   d. Researchers have not requested permission to transfer foreign genes into humans.

5. What is the Human Genome Project?
   a. A project to clone large numbers of identical farm animals
   b. A project to transfer genes from humans into chickens, cattle, mice, and fish

c. A project designed to determine the sequence of the entire human DNA makeup—the entire genetic recipe for humans

d. A project designed to grow supercrops such as supertomatoes and gigantic potatoes

Computer Literacy

1. A computer is
    a. a thinking machine.
    b. a machine that manipulates the symbols of information such as numbers and letters.
    c. an all-purpose device that can distinguish the difference between true and false information.
    d. just a superfast adding machine.

2. In carrying out an information-processing task, the computer
    a. is controlled by a set of detailed, step-by-step instructions called a program.
    b. can distinguish reliable and unreliable input data.
    c. cannot maintain data files and retrieve items from that file on request.
    d. can perform any task if we know what keys to hit.

3. What is the difference between hardware and software?
    a. The two terms mean the same thing in computer jargon.
    b. Hardware consists of the machine itself—the electronic and electromechanical devices needed to process information, whereas software consists of the programs that direct the machine.
    c. Hardware includes those programs permanently installed in the machine that cannot be changed during normal operations, whereas software are those programs needed for specific tasks such as word processing or data management.
    d. Hardware is the keyboard and video screen, and software is all of the wiring and circuitry inside the computer.

4. The component that controls and coordinates all of the functions of a computer is called the
    a. random access memory or RAM.
    b. read only memory or ROM.

c. central processing unit or CPU.
d. word processing unit.

5. Computers store, process, and manipulate information using just two symbols, 0 and 1. These symbols are known as
a. bytes and words.
b. binary digits, or bits.
c. the ASCII code.
d. the base-10 or decimal system.

6. On which of the following may computer data be stored?
a. A floppy disk
b. A hard disk
c. A magnetic tape
d. All of the above

Environmental Issues

1. According to theories about the greenhouse effect, the continued emissions of carbon dioxide and other gases will
a. limit the amount of infrared energy entering the Earth's atmosphere.
b. trap heat within the atmosphere causing global warming.
c. balance the cooling effect·caused by the ozone hole.
d. have little effect on the climate because the oceans can absorb all excess solar radiation.

2. The belief that human activities can alter the amount of stratospheric ozone and thus lead to an increase in the amount of harmful ultraviolet radiation reaching the Earth's surface is
a. a scientific theory that has yet to be proven or disproved.
b. accepted by the scientific community, largely because of the recent measured losses of global ozone.
c. a scientific theory based solely on computer modeling with little or no supporting empirical data.
d. a phenomenon limited to the Arctic and Antarctic regions.

3. What is acid rain?
a. Rain with an acidic rating higher than pH 5.6
b. Rain with an acidic rating lower than pH 5.6

   c. Rain with an acidic rating higher than pH 7

   d. None of the above

4. The most effective way to control pollutants that cause acid rain would be to

   a. burn only coal in our power generating plants.

   b. use only electricity for all our power needs.

   c. burn only petroleum in our power-generating plants.

   d. reduce emissions from all fossil fuel-burning sources.

5. What is the only perfect method for the disposal of toxic waste?

   a. Burning it

   b. Burying it

   c. Dumping it at sea

   d. There is no perfect way

6. The current plan for the disposal of nuclear waste is to

   a. deposit the waste in the Antarctic.

   b. bury it under the ocean.

   c. bury it in an underground depository located in Nevada.

   d. deposit it on a remote island in the Pacific Ocean.

Energy Issues

1. The United States obtains the largest percentage of its energy today from

   a. natural gas.

   b. coal.

   c. petroleum.

   d. geothermal, hydro, and nuclear power.

2. A Btu (British thermal unit) is defined as

   a. equivalent to a barrel of crude oil.

   b. the amount of energy required to raise 1 pound of water by 1 degree Fahrenheit.

   c. equivalent to a gallon of gasoline.

   d. equivalent to a kilowatt-hour.

3. Which of the following is not considered a viable alternative to our continued dependency on fossil fuels until at least 2020?

   a. Nuclear power

   b. Solar power

     c. Hydrogen fuel
     d. Improved energy efficiency

4. What is the difference between nuclear fission and nuclear fusion?
     a. Fission involves the splitting of atomic nuclei into smaller parts, whereas fusion involves the combining or joining of two atomic nuclei.
     b. Fusion is the system used in today's atomic power plants; fission is the process that makes the sun and the stars burn and powers the hydrogen or thermonuclear bomb.
     c. Fission requires very high temperatures to fuse together two nuclei, whereas fusion involves the splitting apart of a nucleus.
     d. Fission reactors use plutonium; fusion reactors use uranium, which is considered much more dangerous.

5. Most applications of solar power today
     a. are cost competitive with other sources of energy.
     b. involve the conversion of solar energy to heat or to electricity.
     c. face only political hurdles and involve few technical or financial constraints.
     d. can produce electrical energy at no cost.

Superconductivity

1. Electricity is the flow of electrons, and the material through which the electrons flow is called a
     a. resistor.
     b. capacitor.
     c. transformer.
     d. conductor.

2. The phenomenon of superconductivity occurs when a material is
     a. heated to a very high temperature.
     b. cooled to a very low temperature.
     c. placed in a vacuum.
     d. in outer space.

3. A material is considered a superconductor when it
   a. becomes magnetized.
   b. loses all resistance to electron flow.
   c. becomes radioactive.
   d. will no longer conduct electricity.

4. A breakthrough in superconductivity occurred when a material was discovered that would lose all resistance at
   a. temperatures above 77 degrees Kelvin.
   b. temperatures below 77 degrees Kelvin.
   c. room temperature.
   d. absolute zero.

5. Superconductivity has great potential for improving
   a. power generation and distribution.
   b. advanced electronic systems.
   c. transportation systems including levitating trains.
   d. all of the above.

High-Technology Medicine

1. CAT scanners represent a major improvement over conventional X-ray technology because
   a. they provide a cross-sectional view of the tissues within the human body.
   b. they are less expensive to use than X-ray machines.
   c. they take fewer pictures than X-ray machines.
   d. they preclude the need for surgery.

2. Sonography uses which of these technologies to look within the human body?
   a. Radar
   b. Sonar
   c. Lasers
   d. Sound waves

3. Digital subtraction angiography (DSA) is an imaging technique that
   a. is not used in American hospitals.
   b. is used to provide clear views of flowing blood or its blockage by narrowed vessels.

    c. does not involve the use of X rays.

    d. requires more dye to be injected into the patient than conventional angiograms.

4. PET scans

    a. are the same as CAT scans.

    b. show an organ's shape and structure but not how it is functioning.

    c. offer functional perspectives of biochemistry in living tissue.

    d. are never used in connection with brain or heart troubles.

5. Which of the following statements is true?

    a. Hybrid organs that combine living tissue with synthetic parts are not possible.

    b. Artificial joints can be made out of stainless steel, plastic, or chromium.

    c. Results of mechanical heart transplants to date have been promising.

    d. The technology of artificial limbs for amputees has not changed in fifty years.

6. Which of the following is *not* true of medical laser technology?

    a. It can be used in place of X rays.

    b. It can be used for some types of surgery.

    c. It can be used to halt internal bleeding.

    d. It can be used to vaporize abnormal growths or tumors.

Transportation Technology

1. How do the First and Second Laws of Thermodynamics affect automobiles?

    a. They do not affect modern engine technology.

    b. They set the theoretical limits of engine efficiency.

    c. They apply only to gasoline-powered cars and not to diesel-powered cars.

    d. They apply to both gasoline- and diesel-powered cars but not to electric cars.

2. The aerodynamic drag of an automobile
   a. increases in direct ratio to the speed of the car.
   b. increases with the square of the vehicle speed.
   c. is the same at any speed.
   d. decreases at higher speeds.

3. An airplane flying at Mach 2 is traveling
   a. at over 2000 miles per hour.
   b. at about 1480 miles per hour.
   c. twice the speed of light.
   d. at near the speed of sound.

4. The major advantage of antilock braking systems (ABS) is
   a. their ability to activate a car's braking system even if the driver has fallen asleep.
   b. their ability to permit a car to be steered even under heavy, emergency braking.
   c. their low cost as compared with conventional braking systems.
   d. their ability to detect obstacles on the road ahead and to activate the braking system without the driver taking any action.

5. Which of the following statements is *not* true about automobile air bags?
   a. Air bags are designed to inflate in a serious frontal crash that occurs at speeds over 12 mph.
   b. Air bags often inadvertently inflate, causing accidents.
   c. Air bags have proven their ability to save lives in frontal crashes.
   d. Air bags inflate within 1/25 second—faster than the blink of an eye.

6. The development of fly-by-wire systems means that future commercial aircraft will
   a. not require pilots.
   b. facilitate quick pilot response to turbulence and other changes in flying conditions.
   c. allow aircraft to be piloted from control stations on the ground.
   d. eliminate the need for computers in the control system.

Superweapons and Arms Control

1. Which statement best describes the concept of mutually assured destruction (MAD)?
   a. The employment of nuclear weapons against an enemy's missile sites and weapon control centers
   b. The ability to conduct a first strike so devastating that an enemy cannot strike back in retaliation
   c. The situation in which either of two superpowers can absorb a nuclear first strike and still be capable of counterattacking, causing an unacceptable level of damage to the attacker
   d. The targeting of nuclear forces against an enemy's leadership and command structure

2. Under the triad concept, U.S. strategic forces are divided into the following three components:
   a. short-range, medium-range, and long-range missiles.
   b. air-based, sea-based, and land-based systems.
   c. mobile, fixed-based, and hidden systems.
   d. nuclear forces, conventional forces, and chemical warfare systems.

3. The original purpose of the SDI (Star Wars) system was to
   a. provide an assured capability to retaliate against a surprise attack.
   b. defend the United States against nuclear attack.
   c. move all nuclear warfare to outer space.
   d. limit nuclear attacks to military targets.

4. The lethality of modern warfare may make it necessary to
   a. increase the number of trained humans in battle.
   b. reduce the costs of weapon systems.
   c. increase the number of weapon systems in the arsenal.
   d. remove the human from the system altogether.

5. Which of the following statements about antisubmarine warfare (ASW) is true?
   a. Sound waves cannot penetrate sea water.
   b. Radar is used to detect and track submerged vehicles.

c. Sonar detection equipment uses sound waves.

d. Both active and passive sonars use infrared detection sys-
    tems.

6. Which of the following statements is true?

    a. Current seismic equipment is not sensitive enough to de-
        tect nuclear explosions anywhere in the world.

    b. Nuclear detonations cannot be distinguished from earth-
        quakes.

    c. New seismic technology can detect even small nuclear
        blasts from long distances.

    d. Seismic technology cannot estimate weapon yields with any
        degree of accuracy.

"It used to be that if someone could write his name and read a few passages from the Bible he was considered literate. That's not going to cut it any more."

Bill Honig, California State Superintendent of Public Instruction. Quoted in the *San Francisco Chronicle*, November 17, 1987.

"We all pay a terrible price for technological illiteracy in this society, and far too many of us are technologically ill-informed."

William R. Graham, Director of the Office of Science and Technology Policy; Science Advisor to President Reagan.

1

SPACE TECHNOLOGY

SPACE TRAVEL, exploration, and the prospect of space colonization fire our imagination. We live in an exciting time in this latter twentieth century, a time of humankind's first tentative steps into space. Most of us are relegated to the role of earthbound observer of this grand adventure, but we can enhance our vicarious pleasure by knowing a bit about the territory in question.

Most of us have had courses in astronomy or have heard Carl Sagan lecture about cosmology on television, but there are at least two important reasons we may still want to sign up for this refresher course: One, we may have been sitting in the back of the class or missed a lecture or two. Two, many new and exciting discoveries have been made in the last few years, and the news may not have reached all of us yet. Do we armchair space explorers really know as much about the problems and challenges of space as we should? For instance, is Venus the hot planet, or is that Saturn? Have we explored Uranus or Pluto yet, and if so, what did we find out? Which moon in our universe has an atmosphere? Where in the solar system is a day longer than a year? According to what we know today, are aliens, UFOs, or extraterrestrial life likely or probable? In the process of answering these questions (and raising many more) we will start first with a review of what's out there, followed by a brief course in basic space physics.

WHAT'S OUT THERE?

Stars: celestial bodies that generate energy by means of nuclear fusion at their core. The Sun is a star.

Planets: celestial bodies more massive than asteroids but less massive than stars that revolve around a star and shine by reflected light. Our solar system contains nine planets and their moons.

Moons: natural satellites orbiting some planets. Earth has one moon.

Asteroids: relatively small solid objects that orbit the Sun and shine by reflected light. Though they number in the millions, their total mass is but a small fraction of Earth's.

Comets: celestial bodies moving around the Sun in highly eccentric orbit. Thought to be lumps of dirt and ice, they often form glowing tails visible in the skies above Earth.

Galaxies: large systems of stars held together by gravitational forces. Our Sun belongs to a galaxy called the Milky Way, which is one of billions of galaxies in the universe.

Quasars: strong emitters of radio energy thought to be too large to be stars and too small to be galaxies. However, they have an energy output greater than 100 galaxies. Thought to be young galaxies billions of light years away.

Pulsars: pulsating sources of intense energy. They are also called *neutron stars* and are formed when massive stars run out of nuclear fuel and collapse.

Black holes: objects in space with gravitational fields so intense that nothing, not even light, can escape.

SPACE PHYSICS FOR ARMCHAIR ASTRONAUTS

Don't let the term *physics* scare you. There is only one equation in this entire chapter and that's one you already know.

Space, from our point of view on Earth, can be divided into three main divisions. The first and by far the smallest of these is our solar system, shown in Figure 1–1. Our Sun, with a diameter of 864,000 miles and its strong force of gravitation, holds the 9 known planets in their elliptical orbits. In addition, our solar system includes 37 satellites (moons) of the planets. This is, of course, not counting the many

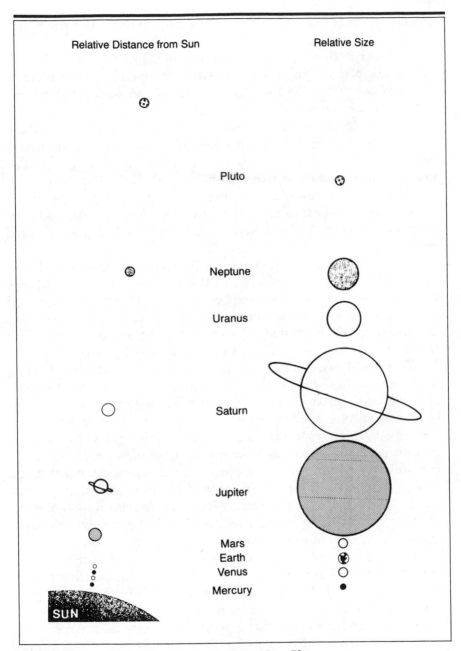

Relative Distance from Sun   Relative Size

Pluto

Neptune

Uranus

Saturn

Jupiter

Mars
Earth
Venus
Mercury

SUN

FIGURE 1–1 *Our Solar System and Its Nine Planets*

artificial satellites that we have sent to space ourselves. The size of this "small" solar system is about 7.3 billion miles across.

Humans in their egocentricity once thought that the entire universe—our solar system plus all the stars—revolved around the Earth. We now know, of course, that Earth is one smallish planet of a medium-sized star on the outskirts of only one of uncounted billions of galaxies.

That brings us to the next major division of space—"our" galaxy. Here we have an aggregation of about 100 billion stars. Our Sun is but an average size star in this galaxy that we call the Milky Way. The nearest star to us after our own Sun is called Alpha Centauri and is so distant that it takes light, traveling at 186,000 miles per second, 4.5 years to reach us. Our galaxy is so vast that it takes light 100,000 years to travel across it. And, I remind you, ours is only a medium-sized galaxy.

The third and last division of space is everything beyond our Milky Way—all the rest of the universe. In the almost unimaginable reaches of outer space are countless numbers of aggregations of suns (called galaxies). They are rotating and moving away from us (and each other) at various speeds calculated as fractions of the speed of light. The further away from us, the faster the galaxies are receding.

This is evidence that the universe is expanding. And it is expanding uniformly. The greater the distance from Earth, the greater the speed of the receding galaxy. A galaxy that is 1 billion light years away recedes half as fast as one 2 billion light years away. In a specific given time, every galaxy increases its distance from every other galaxy by the same percentage. This means that from our point of view on planet Earth, the Milky Way seems to be the center of the expansion and that any other galaxy, from the point of view of its own inhabitants, if any, would seem equally central. The analogy has often been used of a loaf of raisin bread baking in the oven. As the loaf expands each raisin (galaxy) recedes from every other raisin at a uniform rate.

How big is space? In 1987 scientists found what they then considered the farthest known object in the universe: a quasar that may be about 81 billion trillion miles away. A *quasar* is a source of radio energy something like a star (hence the name *quasistellar* or quasar). A billion trillion is a number so large that it strains our ability to comprehend it.

A word is in order here about the electromagnetic spectrum and how we detect sources of energy from space. Electromagnetic energy is generated by natural processes across a wide range of wavelengths, including gamma rays and X rays, visible light, microwaves, and radio waves. We can detect some of this radiation here on Earth, notably visible light and radio waves. The rest, however, is absorbed by the atmosphere and can only be detected by instrument-carrying rockets or satellites in space. Light from the distant quasar would take 13.8 billion years to reach Earth. In 1988, astronomers discovered Galaxy 4C41.17. The most distant galaxy ever detected, it is about 15 billion light years away from Earth.

As you saw earlier, space can be divided into three main divisions, and their relative sizes involved are important. The size of the solar system is 0.001 light years, our galaxy is 100,000 light years, and the universe is about 15 billion light years.

The Speed of Light

The time has come to define a light year because that is the unit of measurement used in astronomy and cosmology. A light year is a measurement of distance, not time. It is used because it is an appropriate measure for extremely large distances. Here on Earth we use miles or kilometers for distances. For example, it would be awkward and inappropriate (but correct) to say that the distance between San Francisco and Los Angeles is 24,520,320 inches. In space, earthbound units of measurements are just too small and lead to numbers with an awful lot of zeros.

A light year is the distance that light, moving at slightly more than 186,000 miles per second, travels in a year. Just for the fun of it, we can convert this to miles just to illustrate how big the numbers get.

186,000	miles per second
× 60 =	
11,160,000	miles per minute
× 60 =	
669,600,000	miles per hour
× 24 =	
16,070,400,000	miles per day
× 365 =	
5,865,696,000,000	miles per year

Because numbers of this size are cumbersome, mathematicians use a shorthand in which exponents (squares, cubes, and so on) express these large numbers concisely. The number 1000 is expressed as 10^3, ($10 \times 10 \times 10 = 10^3$), and 10,000 is expressed as 10^4. A light year then, rounded off, is about 6 trillion miles and can be expressed in miles as 6×10^{12}. A light year, as Figure 1-2 shows graphically, is the speed limit in the universe.

It is time now to turn to the question of how and when our universe began.

Beginning and Endings

There used to be several theories about the origin of the universe. Now, however, the "Big Bang" theory has come to be accepted and is in fact called "the standard model." Several good books have been written on this subject. Probably the best of these and the one most often quoted is Steven Weinberg's *The First Three Minutes* (New York, Basic Books, 1977). His is an exciting book, and I recommend it.

The Big Bang theory states that the universe began when a single point of infinitely dense and hot matter (called a *singularity*) exploded spontaneously. This gigantic explosion was not an explosion of the type we know, starting from a center and spreading out in all directions. The original explosion at time zero occurred simultaneously everywhere, filling all space from the beginning, with every particle of matter rushing apart from every other particle. It was not a burst of matter into space but an explosion of space itself.

The temperature of the universe a few fractions of a second after the explosion has been calculated to have been about a hundred thousand million (10^{11}) degrees Centigrade. This is very hot indeed, hotter than the temperature of the hottest star. This temperature is so hot that none of the components of ordinary matter—molecules or atoms—could have held together. The matter exploding in these early microseconds consisted of various types of what are called elementary particles: electrons, positrons, neutrinos, and photons. Modern high-energy nuclear physics deals with the study of these particles.

As the explosion continued, protons and neutrons were formed. When the temperature dropped until it was cool enough (about 1000 million degrees Centigrade) the protons and neutrons began to form into the complex nuclei of hydrogen and helium.

After a few hundred thousand years, the matter became cool enough for electrons to join with the nuclei to form atoms of hydrogen

FIGURE 1-2 *Universal Speed Limit*

and helium. Under the influence of gravitation, the hydrogen and helium gas formed clumps. These clumps ultimately condensed to form the galaxies and stars of the present universe.

An analogy often used to explain the evolution of the universe is to compare this process to the changes that occur to a cloud of highly compressed steam when it is released. On release, the cloud of steam will expand and cool. When it cools down to 213.8 degrees Fahrenheit (101 degrees Centigrade), it condenses into droplets of water. If it continues to expand and cool, it will reach another important temperature border—32 degrees Fahrenheit (0 degrees Centigrade)—at which point the water freezes into ice. This analogy holds for the universe, except that the transitions corresponding to condensation and freezing were both far more numerous and far more complicated.

In passing we can note two laws of physics implicit in the description of the Big Bang:

1. Compressing matter always makes the molecules in matter heat up.

2. Expanding matter always cools down.

All this occurred, according to the best scientific estimates, about 15 billion years ago. Galaxies formed some 10 billion years ago, and our own Sun and solar system emerged some 5 billion years ago. Earth took shape about 4.5 billion years ago. (Later in this chapter you'll see how these estimates are made and how other theories about the universe are derived.)

There you have the standard model of the beginning. There is no standard model or accepted script for the end. The universe may go on expanding forever, getting colder, emptier, and deader. This is called the *open universe theory*. On the other hand, it may someday collapse into an apocalyptic Big Crunch, breaking down the galaxies, stars, atoms, and atomic nuclei into their original constituents. This is called the *closed universe theory*. A third theory postulates a universe continuously oscillating: Big Bang, expansion, collapse, then another Big Bang. Whatever future is in store, the planet we all know and love will not be part of the picture.

Sadly, Earth's fate is sealed. According to the astronomers, about 5 billion years from now our Sun, having reached its old age, will expand into a red giant and swallow up the planets Mercury, and Venus, and then Earth in a huge fireball. A red giant is an intermediate stage in the life cycle of a star from its birth in a cloud of gas, to its eruption into a red giant, and to its eventual fade into obscurity as a white dwarf.*

It is possible that the human species, if it has not blown itself up or polluted itself out of existence in the meantime, may, by this far-

---

*As a star runs low on fuel, its color changes from yellow-white to deepening red. Ultimately, the stellar atmosphere burns away, leaving behind a dense sphere about the size of Earth—called a *white dwarf* star. Not all stars have the same fate. How stars die depends principally on their mass. Stars at least 10 times the mass of our Sun may culminate in a huge explosion called a *supernova* and eventually become black holes. Medium-size stars, one to three times the mass of the Sun, may evolve into neutron stars or pulsars.

off time, manage to colonize a new planet. If so, our descendants will record the end of the late great planet Earth with, we hope, some melancholy.

How Do Astronomers Know All That?

Unless you want to take my word for all of the foregoing, and the true scientific approach is never to take anybody's word for anything, it's time to review, however briefly, just how the scientists arrived at their picture of the universe.

Astronomers tell us many things about the stars in the universe: they can tell us how far away a specific star is, how big it is, how old it is, and the chemical makeup of any particular star. They do this by using a set of tools: optical telescopes, radio and X-ray telescopes, infrared telescopes, gamma-ray detectors, and most important of all, the techniques for deciphering components of light. These techniques for analyzing the spectrum of light—the visible portion of the electromagnetic spectrum—come under the heading of *spectroscopy*.

A *spectroscope* is an instrument that splits the white light given off by the stars into the spectrum of its component colors from violet through blue, green, yellow, and orange to red. This spectrum is not continuous, like a rainbow, but is interrupted by a number of dark lines. From the relative number, thickness, and position of these lines, astronomers can calculate the temperature and composition of a star, the speed and direction of its motion, its rate of rotation, and the strength of its magnetic field.

Before there was spectroscopy, however, there was trigonometry. A relatively simple and straightforward method for measuring the distance to a star is by what is called parallax. As you can see in Figure 1-3, parallax is a measure of the distance an object seems to move in relation to its background when it is viewed from two different places. Astronomers have measured the distance to all of our nearest stars by viewing them during opposite seasons of the year. This is when half of a complete revolution of the Earth around the Sun has provided a known baseline 186 million miles long.

Now, how about the distant stars, and what about the size, age, and chemical makeup of far-off parts of the universe? That brings us back to light and spectroscopy. When light is focused and made to pass through a prism, we see it spread into a rainbow of colors, which

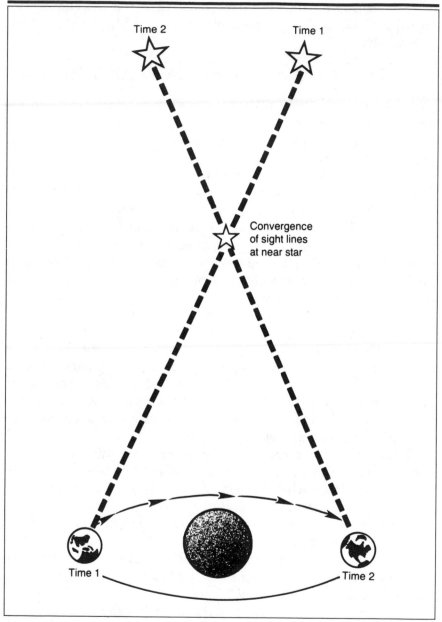

FIGURE 1–3 *Parallax: The Apparent Displacement in the Position of a Star or Planet When Viewed from Two Different Locations, Creating a Means to Determine the Distance to the Star or Planet*

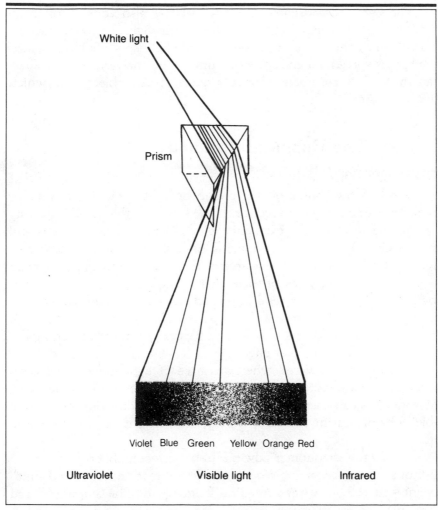

FIGURE 1–4 *Spectrum of Light When Passed Through a Prism*

we call the spectrum of light. Figure 1–4 shows such a spectrum. We perceive different colors because the various light waves are of different lengths. Shorter waves are perceived as bluish colors, the longer wavelengths as reddish hues.

Each type of atom, when it is heated, emits a characteristic glow, consisting of a blend of specific colors. These colors (or wavelengths) are uniquely different for each element.

The light from distant stars and galaxies also consists of many different colors blended together. When this light is passed through a prism (or by the use of more sophisticated light splitting instruments) the light is spread out and the spectrum can be analyzed. Astronomers can then tell what atomic elements make up the object from which the light came.

The Doppler Effect and Red Shift

The Doppler effect is the change in wavelength of sound or light that occurs when the source of that sound or light is moving toward or away from an observer or receiver. If the source of the waves is moving toward the receiver, the frequency of the wavelengths increases and the wavelength is shorter, producing high-pitched sounds and bluish light (called *blue shift*). If the source of the waves is moving away from a receiver, the frequency of wavelengths decreases, sound is pitched lower, and light appears reddish (called *red shift*). A commonly cited example of Doppler effect is the apparent change in pitch of a siren as a fire engine approaches and then moves away from a listener. Figure 1–5 illustrates the Doppler effect.

As with sound waves, when a source of light moves toward an observer, the waves are squeezed together and the colors become bluer. Astronomers can therefore gauge the motion or direction of celestial objects by measuring the color shift in the patterns of color they give off.

In 1929 the astronomer Edwin Hubble detected that the light from distant stars was becoming redder. He determined that this "red shift" meant that the stars were receding from Earth. The amount of red shift increases as the speed of recession increases. And the speed of recession is found to increase proportionally with distance—at about 15 kilometers per second per million light years. This is known as *Hubble's Constant.*

Red shift indicates that all distant galaxies are moving away from us and the more distant the galaxy, the faster its movement away. Before we undertake a quick exploratory tour of the universe, there are two important concepts of Einstein's to consider. The first deals with curved space and the second with a rather mind-boggling idea called *spacetime.*

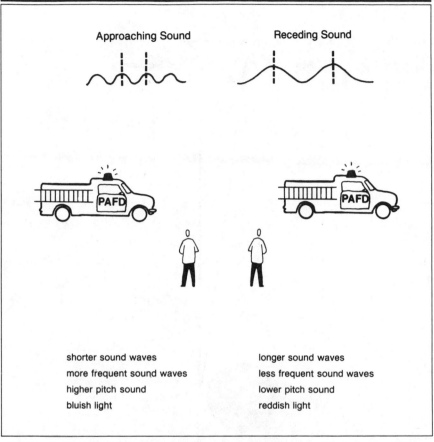

FIGURE 1–5 *Doppler Effect: The Apparent Change in Pitch as the Source of Sound Approaches and Then Recedes*

The Least Anyone Should Have to Know About Curved Space

The essence of Albert Einstein's General Theory of Relativity (published in 1916) is that the presence of matter distorts space and makes it curve. Experiments carried out in 1919 proved Einstein's theory, which had predicted that light waves would bend when they pass close to a gravitational field. Figure 1-6 illustrates how light from a star passing close to the Sun's gravitational field bends closer to the Sun and gives us a false or apparent position for the star.

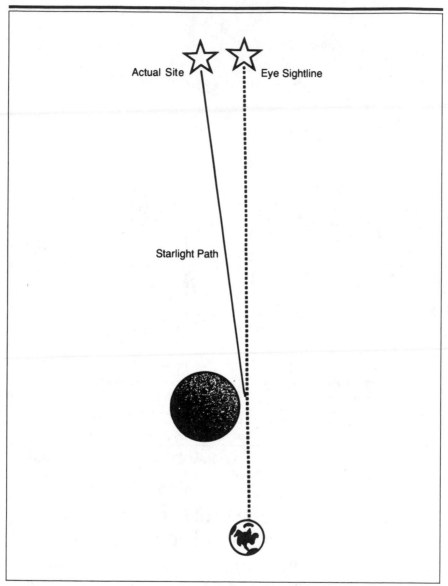

FIGURE 1–6 *The Sun's Gravity Deflecting Light from a Distant Star, Causing the Star to Appear in a Different Position to an Observer on Earth*

Einstein also speculated that radio waves would be bent by distant cosmic objects, and recent experiments have verified that prediction.

To picture Einstein's curved space, imagine a rubber sheet held taut at the edges. If a heavy weight such as a bowling ball is placed on the sheet, it forms a depression in the area of the weight. If you now shoot a marble across the sheet, it tends to curve toward the depression. Now imagine a number of bowling balls on the sheet and a marble traveling with considerable momentum. The rolling marble will tend to curve toward the depressions formed by the weights. In fact, if the depression is deep enough (that is, if gravity is strong enough) the marble is captured by the depression and circles around and around the weight. In general relativity, a large object such as the Sun distorts the fabric of space in just the same way that the bowling ball distorts the sheet. According to Einstein, the large object (mass) in space also distorts time. Let's see how the good doctor explains that.

Spacetime and Dr. E.

In his Special Theory of Relativity published in 1905, Einstein postulated a number of what are now accepted principles of science in space. The equivalence of mass and energy, the relativity of time or the twin paradox that states that a twin who undertakes a long space journey at high velocities would be much younger upon return to Earth than the twin who stayed home, and the increase of mass with speed were among the predictions of the special theory. Let's take these slowly, one at a time, because they do stretch our imaginations.

Begin with the most famous equation in the world and see what it means.

$$E = mc^2$$

In this equation, e represents energy in ergs (a unit of energy), m represents mass in grams, and c represents the speed of light in centimeters per second. Because light travels at 30 billion centimeters per second, the value of c^2 is 900 billion billion centimeters per second. It follows then that the conversion of 1 gram of mass to energy will produce 900 billion billion ergs. An erg is a small amount of energy, and it is hard to imagine. But there is a lot of energy in 1 gram of mass. For instance, the conversion of 1 gram of mass into energy would yield as much as burning 2,000 tons of gasoline.

The important fact to remember is that mass and energy are related and in theory can be converted from one to the other. As the bomb that devastated Hiroshima in 1945 proved, a very small amount of mass can be converted into an immense amount of energy.

Time is the fourth dimension in Einstein's view of the universe. He so mingles time and space in his theory that either concept by itself becomes meaningless. According to Einstein, a clock in motion keeps time more slowly than one that is stationary. He goes even further than this and postulates that all phenomena that change with time (such as aging) change more slowly when moving than when at rest. This is the same thing as saying time itself slows with speed. At the relatively slow speeds at which we travel today, the effect is negligible, but at speeds approaching the speed of light, time slows down appreciably. At the speed of light, time would stand still.

This concept obviously affects future space travel. We travel today, even with rocket propulsion, only at very small fractions of the speed of light. If in some far-off future space travelers were to move at speeds approaching the speed of light, their rate of time passage would be much slower than that for those of us who waited on Earth. They could reach some distant destination and return to Earth in what seemed to them to be only a few years, though on Earth many centuries would have passed. They would return to a world of the future, and their children and their grandchildren would be long dead. (This is a mind-boggling concept, one that science fiction writers have had fun with for a long time.)

Given all of this, it seems that we have to accept the speed of light as the speed limit of the universe. Einstein goes on to tell us still one more difficult-to-comprehend theory. Not only does time slow down with increased speed, but mass increases. A body in motion, he tells us, increases in mass as it increases in speed until, at the speed of light, mass becomes infinite.

If this is indeed true, why doesn't your car get bigger when you step on the accelerator? The answer is that the effect is only significant for objects moving near the speed of light. At 10 percent of the speed of light (18,600 miles per second) mass would only increase by 0.5 percent. At 60 mph, your car is traveling at a relatively slow 1/60 mile per second.

The concept of increasing mass with speed has been well demonstrated in particle accelerators. As the particles move faster, they increase in mass. In fact, the theory is verified every time an accelerator propels particles to very fast speeds. At the Stanford Linear Accelerator, for instance, particles are accelerated to the speed of light in the first few inches, then they simply pick up energy (and mass) and no more speed. To summarize Dr. Einstein:

At the speed of light, time stands still.
At the speed of light, mass is infinite.

Having mastered material in this brief primer, you're ready to do a little space traveling. You don't have to don a space suit, because the vehicle is your imagination. A good place to start is our solar system.

EXPLORING OUR SOLAR SYSTEM

We can start here with what we have accomplished so far, what we are planning in the near future, and what we have learned to date.

As of this writing unmanned U.S. spacecraft have explored every planet in this universe except Pluto, the planet farthest away from the Sun. The Soviets have reached only Venus and Mars. The following Space Log summarizes the major events on the road to space.

Space Log

1957 Beginning of the Space Age, with Sputnik I.

1959 First probes of the Moon, the Russian Luniks.

1961 First man in space: Yuri Gagarin.

1962 First successful planetary probe: Mariner 2 to Venus.

1964 First close-range pictures of the Moon: U.S. Ranger 7.

1965 First successful Mars probe: Mariner 4.

1966 First successful soft landing on the Moon: Luna 9.

1968 First successful manned flight around the Moon: Apollo 8.

1969 First human on the Moon: Armstrong and Aldrin in Apollo 11.

1971 First Mars orbiter: Mariner 9.

1973 First space station established: U.S. Skylab.

First probe encounter with Jupiter: U.S. Pioneer 10.

1974 Second encounter with Jupiter: Pioneer 11.

1975 First Soviet-American space docking: Soyuz/Apollo.

1976 First soft landing on Mars: U.S. Viking 1.

1979 First exploration of the moons of Jupiter: Voyager 1 and 2 reveal that Io, the innermost of Jovian moons, is the most volcanically active body yet found in the solar system.

First encounter with Saturn: Pioneer 11 reveals that rings of Saturn are composed of ice-covered rocks.

1980 First close study of Saturn's largest moon, Titan; discovery of six new Saturn moons: Voyager 1.

Two Soviet cosmonauts return to Earth after spending a then-record six months in orbit.

1981 First flight of a reusable spacecraft: U.S. Space Shuttle Columbia.

Flyby of Saturn by Voyager 2.

1983 Soviet Venera 15 and 16 orbit Venus.

1984 First untethered spacewalk: Bruce McCandless and Robert Stewart in the U.S. Space Shuttle Challenger.

1986 Space Shuttle Challenger explodes shortly after launch, killing seven astronauts, five men and two women; a national tragedy and a major setback to the U.S. space program.

Voyager 2 completes flyby of Uranus and sends back pictures of the Uranian moon, Miranda.

1987 Longest space mission to date: Yuri Romanenko on board the Soviet space station Mir—326 days, 12 hours.

1989 The loss of both Soviet spacecraft bound for the Martian moon Phobos is a major setback to the Soviet space effort.

Voyager 2 rendezvous with Neptune.

Launch of two U.S.spacecraft: Magellan to Venus and Galileo to Jupiter.

The Pioneer 10 and 11 spacecraft are the first vehicles of humankind to venture into interstellar space. Each carries a 6-by-9-inch plaque intended to convey some information on the locale and nature of the builders of the spacecraft, as Figure 1-7 shows. A word is in order about the "space race," because to some extent it still exists and influences the space activities of both the United States and the USSR. At this time the Soviets are focusing their activities on Mars, and U.S. participation is minor. In the 1960s, the USSR dropped plans for a manned moon landing when it became apparent that the Apollo as-

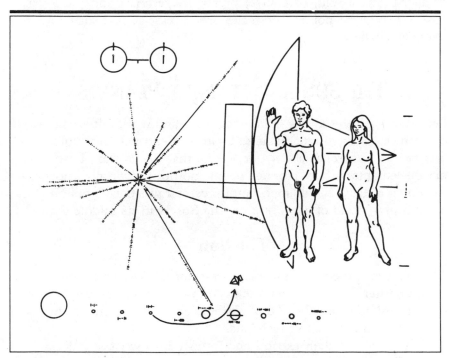

FIGURE 1-7 *Engraved Plaque Carried on Pioneer Spacecraft. Message Indicates Planet from Which Probe Was Launched, Relative Size of Humans, and the Location of the Sun with Respect to 14 Pulsars*

tronauts were going to get there first. Soviets then turned their attention to Mars, but after the successful U.S. Viking probe landed on Mars, the Soviets again canceled a space program because the United States was first. Their next target was Venus; here again, Soviets were influenced by U.S. plans.

When, in the 1970s, we were designing a sophisticated spacecraft to radar-map the surface of Venus, the Soviets reportedly mothballed their less-complicated vehicles, which had been scheduled for the same mission. When budget problems caused NASA to postpone the Venus mission, the Soviets took their vehicles out of storage and launched them as the quite successful Venera 15 and 16 missions. The Venera vehicles orbited Venus in 1983 and produced the first high-resolution map of part of that planet. It is obvious that this vacillation on goals was motivated by politics, rather than by scientific interests, but it is also possible that the space race acts as a spur to both nations. Scientists of all nations would, of course, prefer cooperation and joint space missions, but the political obstacles are at least as formidable as the scientific challenges.

THE SUN AND ITS NINE PLANETS

Let's start with the source of life—the Sun—and work outward toward those mysterious distant planets, Uranus, Neptune, and Pluto. The first fact to remember about our Sun is that it is a star. That seems pretty elementary, but "literate" people do not universally know that the Sun is a star. This all-important (to us) star was born 5 billion years ago out of a cloud of gas. Presently the Sun is in its middle age.

The Sun

The Sun is powered by a nuclear process not unlike a hydrogen bomb. After another 5 billion years, the fusion reactions inside the Sun will have deposited so much helium ash in its core that its nuclear furnace will be forced into hotter reactions.

The Sun will then expand enormously into what is called a *red giant*. The expansion will cool the Sun's surface, but because of its increased size, the total heat that it will radiate will be far greater than at present. Mercury, Venus, and Earth will be consumed in the flames.

After 2 billion more years, the Sun will start to shrink and, in its final phase, will become what is known as a *white dwarf*, a stage in which it cools dramatically. After 50 billion years, the Sun will have turned as black and heatless as space. All this scientists tell us as a result of their analysis of the nuclear processes going on in the Sun.

Mercury

Now consider Mercury, the planet closest to the Sun and the one on which the day is longer than an Earth year. Mercury, the smallest of the nine planets, comes within 28 million miles of the Sun in its elliptical orbit and out to 43 million miles at the farthest point. The period of this orbit (its year) is the equivalent of only 58.5 Earth days, and Mercury rotates so slowly that its day is actually longer than its year.

It is not likely that people will ever visit this little planet, as its surface temperatures are too hostile. One side of Mercury receives sunlight for the equivalent of 90 Earth days and then rotates into freezing darkness for 90 Earth days. The temperature on the side facing the Sun reaches 800 degrees Fahrenheit (427 degrees Centigrade).

Venus

The next planet out from the Sun is Venus. In the last few years we have learned much about this mysterious world. The aging U.S. Pioneer spacecraft has been orbiting Venus since December 1978, gathering data from this cloud-shrouded planet and providing scientists with their first close-up look at what had been called Earth's twin. The planet's surface can't be seen by optical telescopes because the atmosphere contains thick clouds of carbon dioxide laden with sulfuric acid.

Pioneer originally carried five probes, which separated from the orbiter and descended into Venus' atmosphere, radioing their data as they sank. Two survived briefly on the surface, but temperature sensors on all five probes burned out about 10 miles above the surface. The Pioneer orbiter will continue its mission until it begins to run out of fuel, slowly loses its orbital velocity, and sinks lower and lower into Venus' superhot atmosphere. By 1992 it will be flying so low within the dense clouds that it will disintegrate.

Using unmanned spacecraft, the Soviet Union has landed on Venus seven times since 1972, the most spectacular being their two Venera spacecraft landings in 1982. Venera 13 and Venera 14 descended safely through the crushing pressure of Venus' atmosphere—89 times that of Earth. The Soviet spacecraft survived Venus' superhot temperature—817 degrees Fahrenheit (436 degrees Centigrade) long enough to transmit photos and data back to Earth, and to scoop up samples of surface material for chemical analysis.

Whereas Venus and Earth are similar in size and density, Venus is an inferno, because of what scientists call a "runaway greenhouse effect," or overheating caused by trapped solar radiation. Gases that play a part in the greenhouse effect on Earth are carbon dioxide—largely from the burning of fossil fuels—methane, chlorofluorocarbons, and nitrous oxides. These gases trap infrared radiation from the Sun that would otherwise escape back into space, and thus cause the surface of the planet to warm. Environmentalists consider the hellish conditions on our sister planet (where zinc melts in the shade) to be a warning to Earth.

Earth

The third planet out from the Sun and the one we know the most about is our own Earth. Venus at 67 million miles from the Sun is too hot and Mars at 141 million miles from the Sun is too cold to sustain life, whereas we sit here in comfort at 93 million miles from the Sun. Recent advances in the study of planetary atmospheres have made it clear that if our planet's orbit around the Sun were only slightly different, all chances for life on Earth would have ended long ago.

From about 4 billion years ago (shortly after Earth's formation) until 2 billion years ago, Earth's gaseous envelope was a mixture of carbon dioxide, methane, and hydrogen. By 2 billion years ago, plant life had altered the mixture through photosynthesis. Many of the earlier gases had been extracted, and much free oxygen had been added. Computer models of this atmosphere show that a small change in Earth's distance from the Sun would have made a drastic difference.

These models show that if Earth had been formed only 5 percent closer to the Sun—at 88.5 million miles instead of 93 million miles—our planet would have become a deadly furnace. If, on the other hand, Earth had been formed only 1 percent farther away from the Sun, a perpetual deep freeze would have resulted. If these computer simulations are correct, our planet's comfortable temperature and life-sustaining atmosphere are products of a very lucky fluke of planetary placement. Some argue that these results indicate that the chances of Earthlike planets existing in other solar systems may be small, and the possibility of other intelligent civilizations in other galaxies may be rare indeed.

Space explorers have made strikingly similar comments about their views of Earth from space: several of these space travelers have noted the fragility of the planet, or what astronaut Michael Collins called "the little blue marble."

Mars

We now turn to a cold, dry planet, swept by hurricane-force dust storms, with an atmosphere one-hundredth as thick as Earth's and almost no oxygen, but a fascinating place nonetheless—the red planet (so-called because that's what it looks like through an optical telescope), Mars. Located 141 million miles from the Sun, Mars is a small planet with a diameter almost exactly half that of Earth and only a tenth as massive as Earth. Its average temperature is a cold −60 degrees Fahrenheit (−16 degrees Centigrade), but if suitably protected, humans could survive this degree of cold. A Martian day lasts a little over 24.5 Earth hours, and its year is equal to 1.9 Earth years. The planet has a weak moonlight from its two moons—Phobos and Deimos.

Mars is a world of extreme interest to space explorers, because it is more dynamic than the Moon and more hospitable than Venus or any other planet. A number of factors make Mars an attractive goal. It is an Earthlike planet with huge mountains and once-active volcanoes. Mars was once wet and warm rather than cold and dry as it is now. It has daily and seasonal weather changes. Except for Venus, it is the closest planet to Earth. Depending on the route chosen, it takes a spaceship about six months to make the trip to Mars.

Exploration of Mars by means of unmanned spacecraft has been extensive. America's Mariner 9 orbiter made the first accurate, full-planet map of Mars in 1971. The first landing on the red planet was accomplished in 1976 when the U.S. Viking 1 and 2 orbiters sent spacecraft to the surface of Mars in search of evidence of life—they didn't find any.

The Soviet Union entered the Mars race with its Phobos probe missions in 1989. Unfortunately, Phobos 1 was lost due to an erroneous computer command that caused the spacecraft to point its antenna in the wrong direction. The Soviet officials lost radio contact with Phobos 2 shortly after the spacecraft reached Mars and only days before it was scheduled to send a pair of landing craft to the surface of the tiny Martian moon Phobos. This double failure constituted a major setback to Soviet space exploration efforts.

The United States is scheduled to reenter the race for Mars in 1992, when the Mars Observer is expected to orbit the planet, scan for water, and collect geological and climatological data. In 1994, the Soviets plan a landing project using a landing craft, a self-propelled rover vehicle, and a balloon-borne camera. This long-term look at Mars may involve international cooperation.

The United States plans to land a vehicle in 2001 capable of returning to Earth the first physical samples of material from Mars. As in the Soviet lander mission, international participation is expected. With all this talk of multinational efforts, why don't the U.S. and Soviet teams join together in their Mars exploration efforts? This is, of course, a political rather than a technical question, but the USSR has extended the invitation, and the possibility for joint missions presently seems quite good.

Landing people on Mars could be accomplished by 2010 or so. By then the United States, the Soviet Union, or both may have in orbit around Mars a manned space station from which habitats capable of sustaining human life would be landed and the colonization of Mars could begin. An editorial in *The New York Times* suggested that the first thing these pioneers could do once they got there would be to make the planet more habitable. The process is called *terraforming*. From bases on a Martian moon, the *Times* suggested, engineers could fabricate giant mirrors to orbit the planet. The mirrors would direct sunlight to melt the Martian polar ice caps. The resulting flow of vapor would thicken the atmosphere, enabling it to trap more heat, like a greenhouse. Genetically engineered plants from Earth could be spread across the planet, creating oxygen. The result, some hundreds and hundreds of years in the future, would be the greening of the red planet. By this time, humans will very likely have polluted Earth to the point where we will need a new garden of Eden.

Jupiter

From little Mars we move on out into space to enormous Jupiter—482 million miles from the Sun. Jupiter's mass is more than twice that of all the other eight planets put together. It has an atmosphere hundreds of miles deep. Despite its monstrous size and mass, Jupiter has a low density, only a quarter that of Earth's. Jupiter has a retinue of 13 satellites—2 of which are bigger than Earth's Moon—and a turbulent atmosphere complete with cyclones and lightning that have fascinated astronomers for years. Jupiter spins so rapidly that its day is only 10 Earth hours long (note, by the way, that as we proceed on out into space away from the Sun, the planets rotate faster and faster). The planet's orbit around the Sun is a leisurely one, so that the Jovian year is 11.9 Earth years long.

Saturn

Now, in our imaginary voyage, we are 888 million miles out from the Sun visiting one of the strangest—and some think the most beautiful—of the planets, Saturn. Saturn owes its beauty to its rings, which encircle the planet's equator. It is an immense, gaseous, half-formed world, 95 times more massive than Earth but only .7 as dense as water. In addition to its rings, Saturn has 10 moonlike satellites. The largest of these is called Titan and is as big as the planet Mercury. Titan is the only satellite in this solar system known to have an atmosphere, albeit a cold one that is full of poisonous methane gas.

Saturn's ring system is composed of bands of minute ice-coated particles. These strange bands are thought to be only a few inches thick and remain one of the awesome mysteries of our galaxy. Voyager 1 spacecraft in 1980 spotted 1000 or more rings or ringlets. One of the current theories is that these rings and ringlets are the crests and troughs of waves rippling out from Saturn.

Voyager 2 photographed Saturn in 1982, before going on its way to Uranus and then Neptune. Figure 1-8 shows the mission's path and dates of flybys. Saturn data from the Voyager 2 flyby showed violent weather, including whirlpool-like spots and high-speed jet streams surrounding the planet.

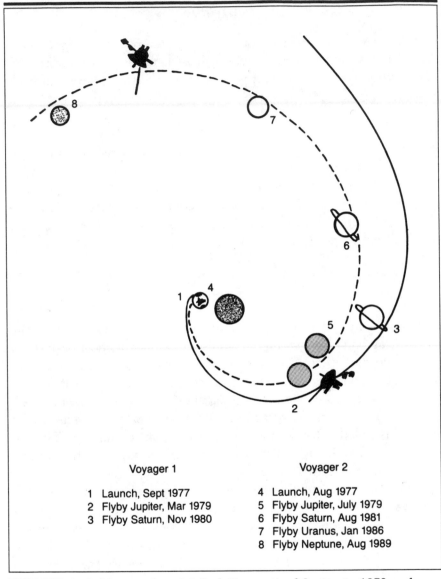

Voyager 1

1 Launch, Sept 1977
2 Flyby Jupiter, Mar 1979
3 Flyby Saturn, Nov 1980

Voyager 2

4 Launch, Aug 1977
5 Flyby Jupiter, July 1979
6 Flyby Saturn, Aug 1981
7 Flyby Uranus, Jan 1986
8 Flyby Neptune, Aug 1989

FIGURE 1–8 *Voyager 1 and 2 Both Encountered Jupiter in 1979 and Saturn in 1980/81. Voyager 2 Encountered Uranus in 1986 and Neptune in Mid-1989*

Uranus

Nine hundred million miles farther out into space lies the chilly methane world of Uranus. Uranus is 14.5 times as massive as Earth, with a temperature of about −270 degrees Fahrenheit (−167 degrees Centigrade). Uranus is the third largest planet in our solar system, Jupiter and Saturn being larger. Recent observations have identified nine rings in nearly circular orbit around the planet.

Like Pluto, the most distant planet in our solar system, Uranus is tipped on its axis. Unlike the other planets, it does not rotate around an axis reasonably perpendicular to the plane of its orbit around the Sun, but rather around an axis that lies almost in the plane of its orbit. In other words, Uranus rolls around the Sun like a ball on a circular track. This odd orientation means that day and night on Uranus have no meaning in Earthly terms.

The U.S. spacecraft Voyager 2 made a flyby of Uranus in January of 1986, and although the encounter lasted only a few hours, the data gathered at that time revolutionized our knowledge of this strange planet, its more than 15 moons, its puzzling newly discovered black rings, and its belt of trapped high-energy particles. Scientists theorize that the rings are made of frozen methane, which reacts under solar ultraviolet radiation to produce carbon compounds.

Voyager 2 was able to obtain and relay to Earth remarkable clear photos of the Uranian moons, Oberon and Umbriel. The surface of the five big moons of Uranus contains copious amounts of water, ice, and some indication of organic matter.

Neptune

Neptune is the eighth planet from the Sun and, with a diameter of 30,700 miles, the fourth largest. Neptune is 2.8 billion miles from the Sun. Prior to the Voyager 2 encounter in mid-August 1989, Neptune was a mystery planet about which little was known. Scientists now think that Neptune is composed primarily of hydrogen and helium and that it has a center of a slurry-like mixture of ice and rock surrounded by hydrogen, helium, and methane gases. The outer layer of atmospheric methane gives Neptune a bluish color. Its great dark spot and some variable clouds near it are evidence of a powerful turbulent weather system.

Neptune is now known to have at least three complete rings and quite possibly two more.

Neptune has at least eight moons—six were discovered by Voyager 2. The largest moon, Triton, is one of the strangest moons in the solar system. Its orbit is the reverse direction of the planet's rotation. It is believed to have an atmosphere consisting of methane and possibly nitrogen. Photographic evidence from Voyager 2 seems to indicate that Triton has a strange collection of violently erupting ice volcanoes that spew up plumes of frozen nitrogen crystals nearly 20 miles.

Pluto

We have now reached Pluto, the ninth planet. Pluto is considered the most distant planet from Earth even though its unusually elliptical orbit sometimes carries it inside the orbit of Neptune. Pluto is a tiny sphere about half the size of Earth. Its year is 248 Earth years long, its temperature is -370 degrees Fahrenheit (-223 degrees Centigrade), and it has one large moon called Charon. Some astronomers believe that Pluto is a former satellite moon of Neptune that drifted into a separate orbit of its own.

Recently, scientists have discovered that Pluto is far too dense to be the loosely compacted snowball they once thought it to be. The current theory is that Pluto has a lot of rock in addition to the methane gas and ice that scientists identified previously. As we have noted, Neptune and Uranus are huge, gaseous balls of very low density. No one is sure why Pluto should be so different. Astronomers have concluded that Pluto's thin atmosphere, consisting mostly of methane gas, sometimes reaches so far out that it envelopes Charon completely, a unique relationship between a planet and its moon. But these same astronomers have speculated that most of the time Pluto does not have an atmosphere.

During Pluto's closer orbit of the Sun, warmer temperature may cause the methane ice on its surface to melt, creating a thin atmosphere. But after a few decades, the surface cools and refreezes as the planet moves farther from the Sun, turning the atmosphere into dew.

Pluto completes our brief tour of the solar system. We'll now turn to NASA's agenda for future space exploration, both manned and unmanned.

WHERE DO WE GO FROM HERE?

Unmanned Space Missions

The current U.S. plans for space exploration are ambitious indeed for a nation that had not launched a new mission to other planets for a decade. It is true that the major space event of 1989 was Voyager 2's close-up view of Neptune and its large moon Triton in August, but Voyager 2 was launched way back in 1977. Let's review the new science missions that NASA has on its near-term agenda.

Magellan

Formerly planned for liftoff in 1988, Magellan was launched from the Space Shuttle Atlantis in May of 1989. Its mission is to map the cloud-shrouded surface of Venus by radar. Magellan was the first planet-bound craft to be launched from a Space Shuttle. It should reach Venus in August of 1990.

Once Magellan gets to Venus, it is to go into an orbit in which it will circle the planet once every three hours. With the planet slowly rotating under it, Magellan should be able to gather radar data for 90 percent of the surface and produce images showing features as small as 150 yards across. A prime objective is to learn why Venus, though similar to Earth in size, has evolved into such a hellish place. Magellan should also be able to obtain information about the extent of volcanic activity and perhaps settle debates about whether Venus once had oceans.

Galileo

Now set for launch on October 8, 1989, Galileo will follow a longer path (refer to Figure 1-9) to Jupiter than had been planned for it in 1986. The delayed mission is now scheduled to reach Jupiter in 1995. On the way, Galileo will swing past Venus and Earth in 1990 for gravitational assists to increase its speed. It will first take a close look at an asteroid (Gaspra) in 1991, pass Earth again in 1992 for another boost, fly by a second asteroid (Ida) in 1993, and finally arrive at Jupiter two years later.

There Galileo will spend 22 months, conducting the first direct probe of the giant planet's atmosphere and photographing in detail its four largest moons: Io, Europa, Ganymede, and Callisto.

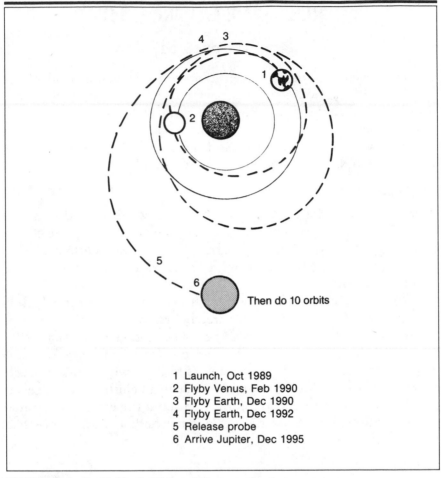

FIGURE 1-9 *Galileo's Planned Route to Jupiter*

Galileo will provide the first direct sampling of Jupiter's atmosphere and the first extended observation of the fifth planet and its moons.

During both Earth flybys, Galileo will use infrared equipment to examine our Moon's dark side, enabling scientists for the first time to map this part of the lunar surface.

Galileo's pictures of Jupiter's moons are expected to have 20 to 100 times better resolution than those taken by Voyager 2 in 1979. After its instruments wear out, Galileo will remain in permanent orbit of Jupiter.

Hubble Space Telescope

The long-awaited Hubble Space Telescope is scheduled to be launched into Earth orbit from the shuttle in March 1990. The Hubble will give astronomers not only a clearer view of the entire solar system, but of stars and galaxies virtually all the way out to the edge of the observable universe.

Astronomers will spend 6 months checking the telescope's functions in orbit from the ground, then as long as 20 years looking at planets and stars with its huge optical telescope. The telescope will provide an unparalleled view, because of its position above Earth's distorting atmosphere. Some scientists believe that the Hubble will bring about a transformation in observational astronomy comparable to Galileo's first telescope nearly four centuries ago.

Ulysses

In October 1990, a spacecraft called Ulysses, built in West Germany, will be launched from Cape Canaveral into orbit around the north and south poles of the Sun. Instruments aboard this international solar-polar mission are being designed and built by scientists from Switzerland, France, Italy, West Germany, and Great Britain, as well as the United States.

Mars Observer

A return to the red planet is scheduled for 1992, when the Mars Observer will be sent into orbit around Mars to study its surface geology and its atmosphere.

Comet Rendezvous Asteroid Flyby (CRAF)

If budgetary approval is obtained, NASA proposes to launch this spacecraft in 1996. It is scheduled to spend five years in space before meeting with the comet Kopff and dropping a probe to its surface. The probe will penetrate to the core of the comet and study the chemistry of its material, which may date from the beginning of the solar system.

Cassini

Planned as a joint mission with the European Space Agency, the Cassini mission will be launched by NASA in 1996, fly by Jupiter in 1999, and arrive in an orbit of the planet Saturn in 2002. Once there, the spacecraft will drop a robot lander to the surface of Titan, one of Saturn's moons, and conduct a chemical analysis.

Scientists believe the thick, organic-rich nitrogen atmosphere on Titan may nurture chemical processes similar to those that were at work on Earth before life developed. The Cassini probe, named after a French-Italian astronomer who discovered several of Saturn's moons in the 17th century, will be launched from the Space Shuttle.

Martian Rover

Before the end of this century, NASA is planning still one more un-manned mission to Mars. The Rover vehicle will be designed to scoop up rock samples and return them to Earth aboard a small rocket launched by radio commands issued from Earth.

Manned Space Programs

The Soviet Mir space station has been in orbit since early 1986 and has been continuously manned during this period. Cosmonaut Yuri Romanenko set a record of 326 days in space while on board Mir. The Soviet Space Shuttle Kosmolyot (Space Flyer) is now in regularly scheduled operation. Although the Soviet space shuttle looks quite similar to the American version, it is in fact a less-sophisticated vehicle. Kosmolyot has no robot arm for retrieving satellites, and the Soviet cosmonauts cannot make space walks from the shuttle untethered as their American counterparts can wearing their manned maneuvering units.

The mission of the Kosmolyot seems to be different from that of the American shuttle. The U.S. vehicle was envisaged as an all-purpose space truck, delivering various payloads into orbit and also able to pick up payloads in space for delivery back to Earth. The main task of the Soviet shuttle is bringing cargo back from space.

Although the success of Mir is impressive, it should be remembered that the U.S. space station Skylab, launched back in 1973, was larger and more sophisticated than Mir is today. Skylab was abandoned in 1974 and allowed to crash back to Earth.

Currently, NASA has plans for a new American space station capable of supporting humans indefinitely, but this project faces some serious budgetary and political hurdles. The NASA plan envisions a space outpost that would cost $8 billion, take eight years to develop and fully construct, and possibly be operational by 1992 (the 500th anniversary of Columbus's discovery of the new world, a point not lost on the supporters of this project).

There is a lot of sentiment among aerospace professionals today that the United States is quite a bit behind the USSR in manned space flight. The United States has accumulated something like 1/10 or less of manned flight time in orbit than the Soviets. On the other hand, there is a body of opinion among many space scientists that NASA is putting too much emphasis on the more glamorous, manned missions at the expense of the unmanned scientific probes. This controversy continues to be aired in the public press.

Scientists were dismayed, for example, when the space agency could not entertain plans to launch a probe of Halley's comet when it returned to our solar system in 1985–1986. On the other hand, many proponents of manned space flight came to view advocates of instrumented space science and technology application missions as chronic complainers, unwilling to support and participate as "team players" in the manned space program.

For now, the manned versus unmanned conflict at NASA continues. President Bush has approved the space station program, and the bulk of NASA resources are to be invested in manned space facilities. Critics claim that the large space station appears to be of limited value to space science and of questionable value as an applications platform for commercial manufacturing.

Important voices in the scientific community have begun to call on Congress and the Office of Management and Budget to "fence off" at least 20 percent of the NASA budget, reserving it by law for space science missions. If this reasoning prevails, we may end up with two separate space exploration programs in the future: a "Scientific NASA" and a "Thrill-a-Minute NASA."

After the space station, plans for future manned exploration missions seem to be centered on Mars. Soviet space planners have talked about a manned mission to Mars by 2010, and NASA too seems to have also selected Mars in its long-term planning for manned missions. There is also the good possibility of a joint U.S./USSR mission to Mars (if politics permit).

You may have noted, astronauts and probes have found no signs of life, intelligent or otherwise, in space explorations to date. Does this mean that we are alone? This is a question that has been asked as long as people have been able to think about the human race and its place in the universe. Is anyone out there? Have they been visiting us in strange crafts from time to time?

THE SEARCH FOR EXTRATERRESTRIAL INTELLIGENCE

The "true believers" take firm positions on both sides of the issue of extraterrestrial intelligent life. Let's try to remain coldly objective in the face of this hot debate. Some scientists, Carl Sagan of Cornell University among them, believe that the probability of extraterrestrial intelligence is overwhelming and that we should undertake a well-planned search for other beings. Critics consider the entire matter unscientific, and any search to be a waste of time and funds.

What about the probabilities? Bernard Oliver, a pioneer in the search for extraterrestrial civilizations, has estimated that there are some 10 billion "good" suns in this galaxy—that is, suns with orbiting planets—and that as many as 1 million of these systems could contain planets like Earth, capable of sustaining life. The conclusion he reached is that the probability of intelligent civilizations is quite good; the problem is contacting them.

The probability of extraterrestrial intelligence (ETI) has been quantified by, among others, the American astronomer Frank Drake. The "Drake Equation" attempts to provide a basis for estimating ETI. Basic to the equation is the assumption that Earth is not special—that is, the conditions that led to the evolution of intelligent life on Earth are not unique and could have occurred elsewhere.

Starting with this basic assumption, a subjective number is then assigned to a number of variables that are considered important to the development of ETI. Among these factors are

- The probable rate at which stars are formed in our galaxy per year

- The probable fraction of those stars that would have planets

- The probability of life forming on a planet

- The probability of intelligent life developing

- The probability that intelligent life would try to communicate with other worlds

As you can see, the equation is completely subjective. If an optimistic number is assigned to all of the variables, the answer will be that the number of possible civilizations trying to communicate with us right now is 200 million. If, however, pessimistic numbers are assigned to the variables, the answer will be 10 or even 1.

Even if we take the optimistic approach, the galaxy is so large that civilizations will be isolated from each other by huge distances, and those distances are growing as galaxies recede from each other. For this reason, scientists consider that even though communication between Earth and ETI is remotely possible, visitation is not—we will see no UFOs.

In the fall of 1988, U.S. and Canadian astronomers found 10 planetlike objects orbiting distant stars, a discovery that intensified speculation about the existence of life beyond Earth. The pro-ETI people took heart with this news, because there was now reason to speculate that as many as half of the stars in the Milky Way may have planetary companions.

Skeptics point out that if advanced civilizations did exist in such great numbers as Sagan and Oliver believe—100,000 or so—then at least one of them ought to have contacted us by now. We know they are not in our solar system, and since there is no evidence for them elsewhere, maybe they don't exist.

Despite the doubts, 72 prominent scientists from 14 countries, including seven Nobel prize winners, signed a petition drawn up by Carl Sagan calling for a search for intelligent life on other planets. In November of 1982, Congress approved a modest $1.5 million in the NASA budget to begin the development of a long-term Search for Extraterrestrial Intelligence (SETI) project.

The SETI search is carried out by means of radio telescopes. The essential features of a radio telescope are a reflector (or many reflectors) to concentrate the incoming radio waves and an antenna, or aerial, placed at the point where the waves are concentrated. The incoming radio waves are then amplified and fed into a recorder that makes a trace of the emission pattern. A computer is used to help analyze the results. As a result of SETI, humans are in effect standing by the "telephone" now, but so far nobody has called.

But why just sit around listening? Could we not plan on traveling to distant solar systems or could not other civilizations travel to us? If there is only one thought to come out of our discussion of the universe it should be the immensity of the distances involved. It takes 4.2 light years to reach the nearest star. At present spacecraft speeds, it would take 40,000 years to reach this star and, of course, 40,000 years to return. You can see why NASA might have a little trouble recruiting astronauts for this journey!

Why can't we travel faster? Here the required size of the rocket is the dominant factor. For example, to escape Earth's gravity field, ships need a speed of about 6 miles per second. This can be achieved with the best-known rocket technology when the rocket weighs about 10 times more at launch than at burnout. However, the weight ratio (the relationship of rocket weight at launch as compared to burnout—usually referred to as "mass ratio") increases *exponentially* with burnout speed. This means that to go faster, say 60 miles per second, ships would need a mass ratio around 10^{13}. To rocket a 200-pound person with an 800-pound spaceship to 60 miles per second would require 1000 trillion pounds (rocket and fuel) at liftoff. The bottom line is that we need improvements in rocket technology by a factor of *billions* before we can even dream about reaching another solar system.

The conclusion to be reached from contemplating these distances and energy requirements is that it is impossible that we will be making any interstellar trips in the foreseeable future. For the same reasons, it is extremely unlikely that UFOs are visiting us every other week (or at all), despite frequent reports in the tabloids.

For now it does appear that we are alone in space as we travel on our little blue marble. There is a positive side to this realization. Knowing we are unique may make us a take a little better care of our home planet.

═══ KEY CONCEPTS ═══

▶ Earth is a small planet orbiting a medium-sized star located toward one edge of an aggregation of stars called the Milky Way galaxy. Earth is the third of nine planets orbiting this star, which we call the Sun.

▶ The Sun is but one of billions of stars in this galaxy and this galaxy is but one of billions of galaxies in the universe.

▶ The universe is expanding, with distant galaxies receding from us and from each other at increasing speeds.

▶ The accepted theory of how the universe began is called the Big Bang, which states that the universe began when a single point of infinitely dense and hot matter exploded spontaneously.

▶ Einstein's concept of the universe states, in part, that the speed of light is finite—nothing can ever move faster, and that

> At the speed of light, time stands still.
> At the speed of light, mass is infinite.

▶ Unmanned U.S. spacecraft have explored every planet in this solar system except Pluto. The Soviets have explored only Venus and Mars.

▶ Plans for future manned exploration missions are now centered on Mars. There is a good possibility of a joint U.S./USSR mission to Mars by the year 2010.

▶ Speculations about the existence of intelligent life beyond Earth remain inconclusive. The 1988 discovery of planetlike objects orbiting distant stars has made it more plausible to assume Earthlike planets with life may exist. However, there has been no evidence of any attempt to communicate nor any real evidence of visitations by UFOs.

2

A Short Course in
BIOTECHNOLOGY
and
GENETIC ENGINEERING

Twice in this century, scientists and their ideas have generated a transformation so profound that it has changed the course of human affairs and has radically altered how we think about ourselves and the world we live in. The first of these transformations was in atomic physics, and the second in biotechnology. The goal of this chapter is to provide a broader insight into this strange world than is normally obtainable from the newspapers or the popular magazines.

THE CIRCULAR LADDER

Over 35 years ago, scientists James Watson and Francis Crick—using cardboard cutouts, wire, and sheet metal models—described the molecular architecture of genetic material. It is the double-railed circular ladder of deoxyribonucleic acid: DNA. What has been learned in the years since has revolutionized the science of biology. Ideas and concepts that have emerged from this revolution touch everyone and have profoundly contributed to our understanding of life itself.

But why should we care about this esoteric and complex field of biotechnology? Those of us who are not biologists or specialists have many other demands on our attention, and time is limited. The answer is simple: We do not have any other intelligent choice. The rapidly growing ability of genetic engineers to redraw life's blueprints has thrust the need to know upon us.

The first order of business is to define the most important terms needed to understand biotechnology:

Amino acids: organic acids that are the building blocks of proteins. About 20 different kinds of amino acids exist in proteins.

Bacteria: single-cell organisms often used as hosts in genetic engineering.

Chromosomes: threadlike structures in a cell nucleus that are made up primarily of DNA. Human cells have a set of 46 chromosomes.

Clone: a cell structure, organism, or group of organisms all derived from a single parent cell and sharing the same heredity.

Cloning: reproducing new organisms from a single body cell, rather than through the combination of sex cells.

DNA: deoxyribonucleic acid, the chemical substance found in the cells of living things that directs the production of proteins and contains genetic information passed on to new cells and new organisms.

Enzymes: proteins that act as catalysts in biological systems—that is, substances that help other chemicals to react. Certain enzymes are used in gene splicing to cut and join pieces of genetic material.

Genes: units of heredity, portions of DNA that direct the production of a specific protein or the expression of a particular trait.

Genetic engineering: the use of scientific, biological techniques to manipulate or rearrange genetic material to alter hereditary traits. Also known as *gene splicing* or *recombinant DNA technology*.

Protein: a complex chemical substance consisting of amino acids that make up the structure of cells.

RNA: ribonucleic acid, a type of nucleic acid that is usually produced as a copy of a portion of the cell's DNA.

Before we review the chronology of development, it is important to understand that biotechnology is not new. Selective breeding of farm animals or boosting crop yields by selecting seeds from the best plants are forms of genetic manipulation that were employed in early civilizations. These traditional methods are still the foundation of modern agriculture. Over the past 50 years, crop yields in the United States have risen steadily by 1 to 2 percent a year. There was, for instance, a steady rise in corn yields in the years 1930 to 1980. Experts have concluded that about 70 percent of this increase was due to genetic improvements resulting from selective breeding.

What is new is that these transformations can now be accomplished far more quickly and precisely through genetic engineering. Genetic engineers now think that within a few years they can achieve the same improvements in yields by genetic manipulation that would have taken decades using conventional plant-breeding techniques. Also new is the application of genetic manipulation to humans. Let's look at the reasons all of this came about so rapidly.

BACKGROUND OF THE BIOTECHNICAL REVOLUTION

The genetic revolution did not start with Watson and Crick and their description of the double helix—the spiral ladder. Dramatic and Nobel prize winning as it was, this discovery was but one important step. You can gain some perspective by checking this brief chronology of developments.

1859 Start with Charles Darwin and his book *The Origin of Species,* which introduced the theory that higher forms of life—including, most disturbingly, *Homo sapiens*—evolved from lower species through an evolutionary process called *natural selection.**

1866 In this year an Austrian monk named Gregor Mendel published *Experiments on Plant Hybrids.* Using garden peas, Mendel demonstrated that traits that are passed from parents to offspring are by means of discrete units called *factors.* These are what we now call *genes*—the basic units of heredity. Mendel's work laid the cornerstone for modern genetics.

---

* The word *theory* in the Theory of Evolution has caused a lot of problems; the thorny issues can be clarified in an aside. We do not speak of the Round Earth Theory because we know from observation and mathematical calculations that the Earth is round. It is a fact. Such is also the case with evolution, because a volume of fossil data collected since Darwin's time demonstrated the fact of evolution beyond rational question. Definitive verification of Darwinian origins came about through the revelations of molecular biology. The word *theory* is therefore a misnomer, because it implies doubt where there is in fact no scientific doubt.

To demand equal time for creation science in biology classes, as some do, makes as much sense as demanding equal time for the Flat Earth Theory in astronomy classes. Or, to put it another way, you might as well claim equal time in sex education classes for the Stork Theory.

1869 Swiss biochemist Johann Friedrich Miescher discovers that the cell nucleus consists largely of a substance whose current name is *nucleic acid*.

1871 A type of nucleic acid called deoxyribonucleic, or DNA, is identified in the sperm of Rhine River trout, but its role in the genetic process remains unclear.

1900 Gregor Mendel's long-forgotten findings are rediscovered and recognized for their importance by three researchers working independently, Hugo de Vries of Holland, Karl Erich Correns of Germany, and Erich von Tschermat of Austria. The study of classical genetics takes on renewed importance.

1944 Three researchers at Rockefeller University—Oswald Avery, Colin McCloud, and Maclyn McCarty—discover DNA to be the carrier of genetic information directing all cellular functions.

1953 James Watson and Francis Crick, working in Cambridge, England, describe the double-helical (double spiral) structure of DNA for the first time.

1970 A class of specialized protein is identified by U.S. microbiologists Daniel Nathans and Hamilton Smith; it can be used to cut DNA strands at precise locations.

1972 A Stanford University team led by Dr. Paul Berg synthesizes the first recombinant (or recombined) DNA molecule by combining DNA fragments from a bacterium and a virus in a test tube.

1973 The experiment that ushers in the age of genetic engineering is performed by Herbert Boyer of the University of California in collaboration with Stanley Cohen of Stanford. They create what they call "DNA chimeras" by splicing genes from a South African toad into a species of bacteria. Chimeras are organisms with the tissues of at least two genetically distinct parents. The new creature reproduces itself, and the offspring carry out the instructions of the inserted genes.

1974 Concerned about the potential of their discoveries and the possible dangers to public health, leading geneticists urge a worldwide moratorium on gene splicing until safety guidelines can be worked out.

1975 Scientists from 19 countries gather at Asilomar, California, to debate the need for safety rules so that research can continue.

1976 The first set of guidelines for federally funded recombinant DNA research is issued by the National Institutes of Health.*

The safety issue of recombinant DNA becomes politicized when the City Council of Cambridge, Massachusetts, becomes concerned about the possible health hazards of research at MIT and Harvard. The council declares a moratorium on DNA research and sets up its own regulatory body.

Genentech, the nation's first biotechnology company, is formed in South San Francisco, California, by researcher Herbert Boyer and businessman Robert Swanson.

1978 A team led by Herbert Boyer produces a human protein—called *somatostatin*—in bacterium as a result of the insertion of a human gene into this bacterium.

1980 The U.S. Supreme Court rules five to four that microorganisms can be patented. The case at issue involves a bacterium that digests oil and can be used to clean up oil spills.

1981 The Cetus Corporation, the nation's second biotech company, goes public and sets a Wall Street record for the largest sum ever raised in an initial public offering.

1982 Sale of the first drug manufactured by recombinant DNA is approved by the U.S. Food and Drug Administration.

---

* Reaction to these guidelines is mixed. Some noted scientists consider them an absurd overreaction. The cautious approach, however, allays the more outlandish fears of both scientists and the public. To date, hundreds of thousands of recombinant DNA experiments have been performed without a single report of a serious genetic accident. Today 90 percent of all recombinant DNA experiments in the United States are exempted from NIH guidelines.

The product is a synthetic insulin produced by Eli Lilly & Co.

1986 Biologics Corp. receives approval from the U.S. Department of Agriculture to market a living genetically engineered organism—a virus used as a vaccine for swine.

The first outdoor field test of a genetically altered crop—tobacco plants engineered to resist disease—is conducted in Wisconsin.

The first set of rules for reviewing and approving biotechnology products is issued by the U.S. government.

1987 The U.S. Patent Office declares it will allow patents on genetically engineered animals—that is, on new species.

1988 The first authorized release of genetically altered bacteria outdoors occurs in Contra Costa County, California, when Frostban™ is used on strawberry plants. Frostban is a recombinant microbe that protects plant leaves against frost for five years.

A small company in Oxford, England, begins selling custom-made genes to pharmaceutical companies. It has 19 genes in its catalog. The custom genes, including several interferons, are intended to be used in the development of new drugs or enzymes to combat specific diseases.

The U.S. Patent Office approves a patent for a genetically altered mouse, the first time that a patent has been issued for an animal. The mouse is to be used in research on breast cancer. Biotechnology critic Jeremy Rifkin opposes the move and claims that the Patent Office is establishing public policy that will have profound economic, environmental, and ethical consequences for the world community. Who needs a new mouse, you say? The new mouse is not the point; critics' reasoning is that if scientists create a new mouse, why not a new elephant or even (organ music rises to a crescendo here) a New Human?

Scientists at MIT, led by biologists Paul Schimmel and Ya-Ming Hou, decipher a second genetic code that has eluded molecular biologists for two decades. The code directs one of several steps in the synthesis of proteins inside cells.

Scientists inject 2200 corn stalks with a microbe that is effective against the destructive European corn borer in the first approved outdoor test of a genetically engineered "plant vaccine."

1989 The U.S. government gives scientists permission to transfer foreign genes into human beings for the first time, approving a test in which bacteria genes are used to track the effectiveness of a new cancer therapy.

With this brief history behind you, we turn now to an even briefer refresher course on the fundamentals of biotechnology.

THE ABCS OF DNA

To begin with, all plants and animals are made up of cells, and every cell—whether in you or me, or an earthworm or a redwood tree—has specific characteristics. Inside each cell is a small, round object set off from the rest of the cell. The object contains the chromosomes and other materials necessary for cell reproduction. This round object is called the *nucleus,* and we'll adjust our imaginary microscope in order to take a closer look at this interesting part of the cell.

Before we do that, however, you'll want to understand one of the terms essential to any discussion of DNA. A *molecule,* from a Latin word meaning small mass, here means the ultimate, individual unit of a substance. For instance, a molecule of water is the ultimate amount of water. It can, of course, be broken down into single atoms of hydrogen and oxygen, but then it is no longer water.

Back to the cell (which is to biology what the atom is to physics). As you know, the cell nucleus contains chromosomes. Each species of plant or animal life has a fixed and characteristic number of chromosomes. Humans, for instance, have 23 pairs or 46 chromosomes. Cells of pea plants have 7 pairs, those of a fruit fly 4 pairs, and bacteria have only a single, unpaired chromosome.

Chromosomes are collections of genes. Genes are the basic unit of heredity. They are the chemical data banks that pass specific traits from one generation to the next. Genes are located on double-stranded molecules of deoxyribonucleic acid, or DNA, which in most organisms are contained in chromosomes. The particular ordering of chemicals in a segment of DNA determines exactly what information is stored in a given gene. The long string of genes that makes up a chromosome has been likened to a string of pearls.

The Spiral Ladder: A Closer Look

If DNA is indeed the key to heredity and the basis of life, how does it work? The DNA molecule as described by Watson and Crick is a three-dimensional, somewhat complicated structure. It can be simplified for our purposes here, as shown in Figure 2-1. Imagine, if you will, a flexible rope ladder that is twisted to form a sort of spiral staircase. The legs of the ladder are formed by units of two different chemicals that alternate along the entire length of the DNA molecule. One of these chemicals is the sugar deoxyribose. The other chemical is phosphate. Each step or rung of the DNA ladder is composed of two chemicals that belong to a group known as *nitrogen bases*. DNA contains four different nitrogen bases: adenine (A), guanine (G), thymine (T), and cytosine (C). The way in which the bases combine to form the rungs of the ladder determines both the structure and the function of DNA.

For any individual of any species, the sequence of base combinations on the ladder spell out a complex coded message that can transmit to an offspring all the instructions needed for every genetic trait. DNA serves as a template from which copies can be made. The coded sequence of base rungs contains a unique pattern setting forth the chemical specifications for some living creature.

The particular sequence and length of the nucleotide chain (the sugar-phosphate-base pattern) in the DNA molecule is what makes a mouse give birth to a mouse instead of a lizard, goldfish, or plant. This DNA molecule determines, among many other traits, the color of our eyes, the pigmentation of our skin, the texture of our hair, and the shape of our head. If all the DNA instructions for a human baby were spelled out in English, it would require several sets of 24-volume encyclopedias. This is a complex concept. (As one scientist put it: Physics is checkers, biology is chess.) Stick with it, though, and reread the chapter if necessary to review what you know so far.

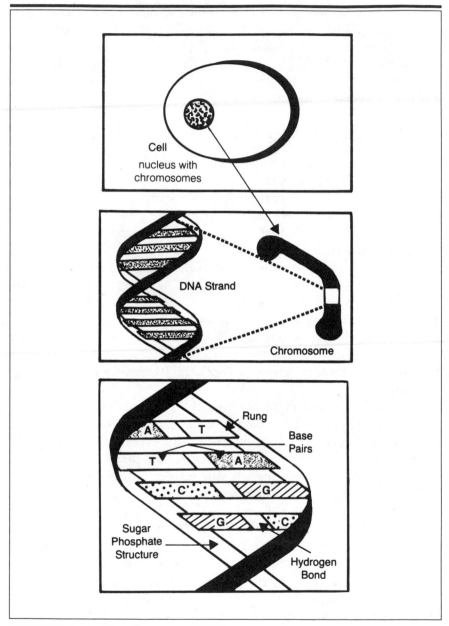

FIGURE 2–1 *The Structure of DNA*

DNA's double helix consists of two intertwined strands, each composed of chains of four different chemical bases (abbreviated A, C, G, and T). Within a strand the bases can be arranged in any order, but the links between the strands can be made only between two specific pairs (A to T and G to C). These bases are so organized as to represent a code to specify the assembly of amino acids into proteins. The code for all 20 amino acids used to make proteins is found to be different groups of three of DNA's bases in sequence. These threesomes are called *codons*. A stretch of codons together forms the instructions for the building of protein. Each stretch is in effect a gene.

Protein Engineering

Scientists are currently working on what is considered to be the next big step in the biotechnology revolution: protein engineering. This involves the creation of complex new compounds custom designed to meet human needs. Called a cross between gene splicing and computer modeling, protein engineering will be used to make entirely new drugs to fight cancer and AIDS.

How important is protein? Virtually all substances produced by living cells are proteins: hormones, enzymes, antibodies, hair, skin, and bones. And virtually all biological functions are controlled by proteins. Being able to tailor-make proteins, therefore, holds great medical promise.

If genes are the blueprints for the body, then proteins are its structural parts—its walls and beams, and the machinery that makes it run.

As you have seen for genetic engineering, the genetic material of cells is altered to tailor-make proteins. A newly deciphered second genetic code plays a key role in a later step in protein synthesis. It provides a second point in the process of protein synthesis at which the interpretation of a cell's genetic instruction can be altered.

The new code, deciphered in 1988 by two MIT scientists, Dr. Paul Schimmel and Dr. Ya-Ming Hou, helps explain a crucial aspect of protein synthesis: the attraction of building blocks, amino acids, to genetic material inside the cell.

In protein synthesis, the DNA code is copied into a similar material called *messenger RNA* (ribonucleic acid). Then the messenger RNA is "read" by other molecules, called *transfer RNA*. These in turn bind to amino acids, assembling them in proper order to make a particular protein. There is a different transfer RNA molecule for each of the 20 types of amino acids found in cells.

Understanding how a particular transfer RNA could cause the binding of a particular amino acid was the challenge for genetic researchers. The second genetic code is the language of the instruction on the transfer RNA that specifies which amino acid will be attracted.

Once researchers understood the DNA/RNA coded message, the inevitable next step was to change the code or engineer the genetics.

DESIGNER GENES

This section is organized into four types of genetic engineering applications, followed by specific examples:

> Producing Drugs in Microorganisms
> > Human Insulin
> > Milk Production Booster for Dairy Cows
>
> Producing New Organisms
> > Mice with Rat Growth Hormone
> > Supertomatoes
> > Engineered Fish
> > Potential for Humans
>
> Identifying Genetic Variations Within a Population
> > DNA Fingerprinting
> > Forensics and Paternity Testing
> > Screening for Genetic Diseases
>
> Cloning of Organisms

Producing Drugs in Microorganisms

Genetic engineering is the use of precise biological techniques to rearrange genes—to remove, add, or transfer them from one organism or location to another. The best known technique to accomplish this is called recombinant, or recombined, DNA. Specialized enzymes are used to snip a gene from one organism and splice it into another. If the transfer is successful, the recipient organism will carry out the instructions of the new gene. The process of splicing a gene is shown for the common bacterium *Escherichia coli* in Figure 2–2.

In the example shown, genes are implanted in microorganisms, such as single-cell bacteria or yeast. The engineered cells become tiny factories, manufacturing the protein ordered by the new gene. The new ability is also passed on to the offspring of the rapidly reproducing cells.

Another example is shown in Figure 2–3. Here biotechnology is used to manufacture human insulin. Bacteria make ideal factories for this process. They are simple, single-cell creatures whose biochemistry is well understood. Bacteria reproduce quickly and they carry *plasmids*, small loops of self-replicating DNA that float around in the cell and are therefore ideal for the insertion of new genes.

The first step in the mass-production of human insulin is to isolate the gene coding for human insulin and cut this gene sequence out of its surrounding DNA by the use of a "restriction" enzyme. These enzymes have proven to be useful tools in genetic engineering. The enzymes had been seen to attack the DNA of invading viruses by cutting it at specific sites, wherever the enzyme found a certain sequence of bases. (To date, over 400 restriction enzymes, capable of cutting DNA at more than 100 different sequences of bases, have been identified.)

Referring to Figure 2–3 again, note that the same restriction enzyme is used to cut open the bacterium's plasmid loop. The human insulin gene is then recombined with the bacterium plasmid and reintroduced into the bacterium. The engineered bacteria multiply in fermentation tanks, producing human insulin.

Supercows will evolve when farmers start using a hormone called BST, which is scheduled to be available in 1990. BST is a recombinant drug that can boost milk production in dairy cattle by 30 percent from about 6 percent more feed. BST is expected to increase milk production per cow by 522 pounds (at least 60 gallons) per year. As a result, the number of dairy cows needed to meet U.S. national needs will drop from about 11 million to about 8 million. The FDA has already approved the sale of milk and meat from BST cows. The BST residue remaining in the milk or meat is broken down by human digestive enzymes and thus is of no danger.

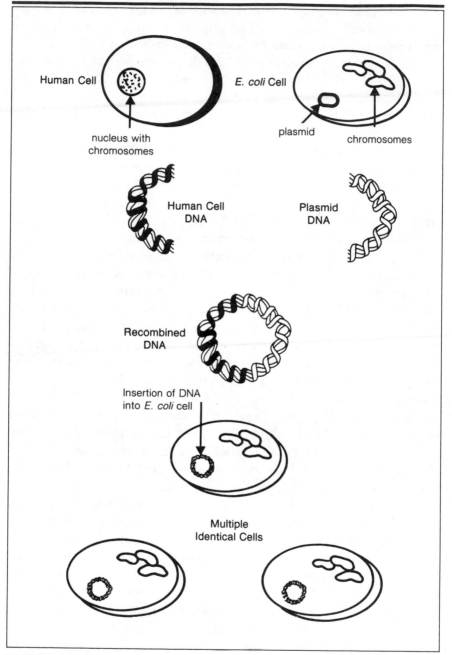

FIGURE 2–2 *Gene Splicing Technology*

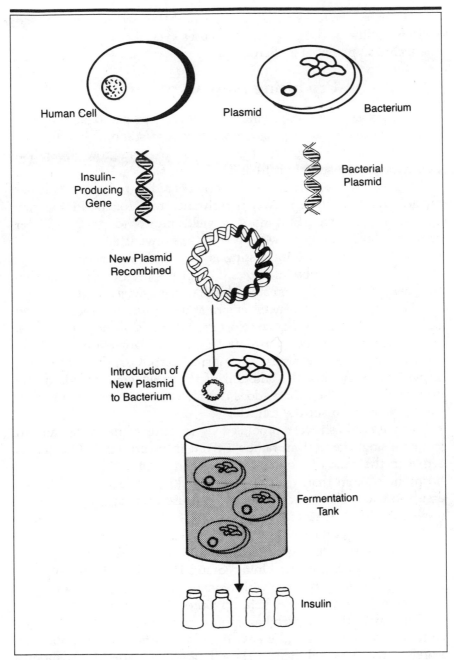

FIGURE 2–3 *Manufacturing Human Insulin*

Of course, we face social/economic problems by making an already efficient industry that produces surpluses even more productive, but the technical potential is there.

Producing New Organisms

Genetic transfers are also being performed with plants and animals. Crops can be enlarged or programmed to grow faster or fight off natural pests. Mice have been given rat genes to make them grow bigger. Two examples are the supertomato and enlarged fish.

Supertomatoes were genetically engineered to resist a specific herbicide used to kill weeds. When resistance to a tomato-killing virus and a hornworm were also inserted, yields improved by 20 to 30 percent and tomato growers' weed control costs went down.

What is even more interesting is that these experiments proved that resistance could be inherited. The plants used in the trials were the grandchildren of the tomatoes that had been genetically engineered.

The application of genetic engineering to fish has recently been demonstrated. A gene that controls growth in rainbow trout was introduced into a common carp, creating a new, larger fish that grows as much as 20 percent faster than normal carp. Carp have their own growth hormone, but the addition of the trout gene speeds up the growth process. The process does not alter the taste or the quality of the edible portions of the fish.

The altered fish were produced by isolating, duplicating, and injecting a single gene from rainbow trout into carp eggs. The gene is active in the cells of trout pituitary glands and produces a growth hormone protein that regulates how quickly the fish grow. Fish are easier to manipulate than mammals, because their eggs are large and grow outside the female.

The potential for this research is enormous. The market for domestic varieties of fish, genetically altered or not, is very large. The consumption of fish in the United States is rising, and domestic production is not keeping pace with demand. In 1988, Americans spent over $6 billion on imported fish, up almost a billion from 1987.

Obviously, this research has important consequences for the aquiculture industry and for the environment. Catfish farmers, and ultimately consumers, will benefit should catfish be raised in 12 months instead of the 18 months it now takes. In addition, other species of fish, including walleye pike and striped bass, could become easier to domesticate and produce on fish farms.

Environmental scientists, however, point out some risks. Scientists are currently attempting to transfer genes from humans, chickens, cattle, and mice into various species of fish. These experiments could substantially change the characteristics of aquatic species. If the genetically altered fish escape from holding ponds, the result could cause ecological disruptions in streams and lakes.

Genetic engineering of higher forms of life is technically more difficult. Although some traits are controlled by a single gene, others reflect the interaction of several genes in a process that is still not completely understood.

Potential for Human Gene Engineering

Now comes the key question: Can scientists engineer human genes? The answer is, not yet. The sad truth is that despite its great potential, gene therapy to correct nature's tragic errors is still consigned to the science of the future. Why is it that not one person has yet to be cured of an inherited genetic defect? Although understanding of genes has grown greatly, not enough is known to practice human gene therapy.

Scientists are currently working with other mammalian creatures to perfect surgical methods of curing certain types of genetic disease. Experiments with human subjects could begin soon and someday, perhaps 5 to 10 years from today, gene therapy will be performed on humans. If human gene therapy has been slower to develop than originally predicted, other advancements in biotechnology have developed in surprising areas.

Identifying Variations Within a Population

Law-enforcement officials are excited by the potential for DNA fingerprinting or genetic marking in forensic science. Police departments around the world have hailed this new technique as the greatest advancement in crime detection since ballistics analysis. The process isolates genetic markers contained in the hair, semen, blood, or skin found at the scene of a crime and compares them with genetic markers of potential suspects. Cruder techniques in the past indicated when the suspect had the same blood type as found on the scene and therefore *could* be guilty. The new technique can indicate with 99 percent probability that the suspect *is* guilty—or at least was at the scene of the crime. Thus, DNA typing can now approach the precision of a fingerprint.

Other applications of DNA identification include paternity testing and prenatal screening. Identification of fetuses with defective genes may lead to corrective treatment or to selective abortion when treatment is not possible.

Screening of individuals for every possible genetic disease, which might be possible in the future, will raise ethical and legal questions. Could an individual, for instance, be denied insurance if he or she carried a "bad" gene? Will insurance companies require genetic screening in the future, or will employers require genetic screening for potential new employees?

Cloning of Organisms

Biological technology now enables livestock breeders to clone large numbers of identical animals from a single embryo. This technique allows the precise duplication of superior animals. The ultimate objective in animal husbandry, that is, achieving the same levels of uniform quality in farm animals that are now confined to manufactured goods, is now possible. How cattle are cloned is shown in Figure 2-4. The process begins by inseminating a prize cow with the thawed semen from a prize bull. This artificial mating yields 10 embryos. Each embryo is composed of 32 cells, each containing the entire package of genetic instructions necessary to form a complete calf.

Scientists, using microsurgical techniques, remove the nuclei from the embryonic cells. They then transfer the nuclei into the cavities of unfertilized eggs taken from ordinary cows and transplant the new embryo into the surrogate mothers. After a normal pregnancy, the surrogate cows give birth to genetically identical calves. The calves produced by this technique are normal in every respect. The offspring represent the state of the art for a technology that is expected to change the way that the finest breeding bulls and cows are produced. Cloning will drastically alter and increase the number of superior animals produced in the $30-billion-a-year beef industry and the $18-billion-per-year dairy industry.

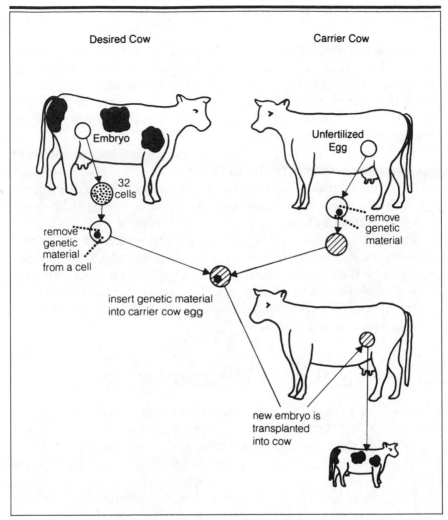

FIGURE 2–4 *How Cattle Are Cloned*

The ability to successfully clone large mammals hints at the possibility in the future that similar techniques might be used on humans. The possibility that a human embryo could be manipulated in the laboratory to produce numerous genetically identical babies carried to term in the wombs of surrogate mothers will add to the controversy about how far we want to go in genetic engineering.

KAMIKAZE GENES

Concern about releasing living gene-altered microbes into the environment has led scientists to program bacteria to destroy themselves when their work is done. This novel technique is based on the discovery that many species of bacteria have "suicide genes" that help to regulate the life and death cycles of their microscopic hosts. When these genes are switched on, the bacteria die.

Controlling this mechanism may prove to be one of the most important advancements in the troubled and controversial campaign by biotech companies to replace toxic pesticides and chemical fertilizers with microbial products. It might even be possible to use suicide genes to attack hazardous chemicals such as PCBs in toxic waste.

The kamikaze genes can be programmed to switch on at different temperatures, light levels, or the presence of specific chemicals. Suicide genes can be the answer to the modern Luddites who fear that new species released in the environment will grow in an uncontrolled manner, to our peril. This fear has been called by one biologist the creature-from-the-black-lagoon syndrome.

THE HUMAN GENOME PROJECT

A major new project has been initiated to map every human gene. This new project is to biotechnology what the Manhattan project was to physics or the Apollo program to space science—a large-scale research program estimated to cost at least $3 billion and take 10 to 20 years.

The project is federally funded and is budgeted at $200 million per year for 15 years. Both the National Institutes of Health and the Department of Energy are involved, and scores of laboratories throughout the United States are carrying on the research.

Called the Human Genome Project, it will be an attempt to determine the sequence of the entire 3 billion base pairs in the human genome DNA that spell out our genetic endowment. A *genome* is the full set of chromosomes—all the inheritable traits of an organism.

Scientists working on this project will attempt to locate, map, and identify all the 100,000 individual genes that are carried in the 46 chromosomes that lie within each of the human body's 10 trillion cells.

Because each gene, in turn, is composed of an estimated 3 billion paired molecules known as *nucleotides,* mapping the entire human genome involves the incredible task of determining the sequence of 3000 trillion nucleotides.

The result of this mammoth research effort would be a list of bases, enough to fill 5000 average-sized books, that would spell out the genetic recipe for humans.

Knowing the full genetic sequences may enable biologists to diagnose and treat both inherited diseases and diseases not now curable, such as AIDS or cancer.

THE BIOTECH CONTROVERSY

Controversy has been a part of the biotechnology picture since the beginning. Although debate has been centered around various issues, the essential question remains the same: Given the power to change nature, how far should we go? For instance, the first applications of genetic engineering were predicted to be in the area of eradicating the grosser genetic misinstructions—that is, the chromosomal mistakes that produce hemophilia, spina bifida, Down's syndrome, or other fearful and tragic forms of human suffering. Attempts to rectify genetic disorders are clearly good objectives. The consequences, however, cannot be ignored. As we gain the power to alter bad genetic instructions, we are also gaining the power to engineer other instructions that people may think good. That is a disturbing idea. What scientist, philosopher, doctor, or politician can we trust with this power—the power, that is, to produce what he or she thinks is the most admirable, valuable, useful human being?

Unraveling the mysteries of genes has already led to the development of new pharmaceuticals, crop plants, livestock varieties, and other products. But this growing capability now confronts us with problems never before imagined.

From the strawberry patches in California—where the first authorized release of engineered organisms was fought and almost sabotaged by protesters—to Washington, D.C., where Congress debates the patentability of life, practical and ethical questions prevail.

Only a few decades ago the term DNA was unknown outside the laboratory. Today, if the results of a recent Gallup poll can be believed, half of the population claims to know what it is. Do they though? This poll revealed a certain amount of confusion. About 80 percent of the public saw biotechnology as a benefit to society, but paradoxically some 70 percent saw that it also brought risks. Three-quarters of respondents believed that the benefits outweighed the risks, and over half thought that unjustified fears of genetic engineering have impeded the development of valuable new drugs. At the same time, three-quarters of the public agreed with the statement that "the potential dangers from genetically altered cells and microbes are so great that strict regulations are necessary." Over half would not allow the approval of a product if its risks were unknown.

Many people today are uneasy about "tampering with life," and this uneasiness translates to demonstrations, protests, and in a democratic society, politics. We are all going to be involved in the decision making, and this involvement will increase as the technology advances. We have learned in other fields that if something is technologically possible, it will happen. There is no use in expecting international consents or self-limiting agreements among scientists to stop biotechnical research. Both have been tried in recent years, and both failed. Technical advancements are inexorable and inevitable.

Society will, however, be required to provide guidelines and limits, and each of us will be involved in structuring the decision-making process. Our choices are going to be among the most important that we will make in the future, so possibly the homework that we have done so far will be worth the effort.

As is often pointed out, there is a similarity between the position of the nuclear physicists 40 years ago and that of biotechnologists today. We are now entering a period of intensive biological research and are considering a wide variety of new applications of genetic engineering. We can only hope that the historical experience of the physicists can provide some guidance to the biologists and enable the genetic engineering industry to avoid the mistakes that trouble the nuclear power industry today.

Thinking about our mind-body system as a collection of programmed chemical instructions has been an extraordinary new way to look at ourselves. Putting environment aside, we generally conclude that we are who we are largely because of our genes. Everything we will be and can be is influenced by our genes and how we interact with the environment. It is difficult for many of us to accept the full significance of this as well as the implications of the coming genetic revolution.

Many scientists now believe that the budding science of genetic engineering will transform the world at least as much as the industrial revolution did. Indeed, some scientists refer to the advent of genetic engineering as "the second Big Bang." If they are correct, it may be important for all of us who consider ourselves technologically literate to keep abreast of developments in this rapidly evolving science.

═══ KEY CONCEPTS ═══

▶ DNA (deoxyribonucleic acid) is a chemical substance that is a vital ingredient in all forms of organic life.

▶ The DNA molecule occurs in the form of an immensely long double spiral, like a twisted rope ladder. The rungs of the ladder form a code and the code contains all the information—the genetic blueprint—describing the creature from whose cell the DNA was taken.

▶ Recombinant DNA technology, also known as gene splicing or genetic engineering, involves taking a portion of the genetic material from one organism and putting it into another organism.

▶ Current applications of genetic engineering include new drugs for humans and animals, new crops, and new fertilizers and pesticides. Future applications include genetically enhanced livestock and gene therapies to repair human genetic defects.

▶ The Human Genome Project is a government-sponsored multi-billion-dollar effort to map all the molecules in all the genes that govern the heredity of the human race.

▶ Biotechnology has the potential to be a major force in our society, influencing the way in which we treat the sick, produce food, and manufacture drugs. Biotechnology also has the potential to alter human genes in a manner that will be transmitted to future generations.

▶ Some people are concerned not only with what they perceive as the hazards to health and environment posed by biotechnology but also with its ethical implications.

=== 3 ===

Babbage to Artificial Intelligence:
COMPUTER
LITERACY

IF YOU CAN DRIVE your car on the freeway without completely understanding the theory of internal combustion engines or use your microwave oven without training in electromagnetic radiation, do you need to understand the inner workings of a computer to operate one? The answer, of course, is that you don't have to become a computer expert or a programmer to run a personal computer. This "user-friendly" chapter is much less ambitious; it describes the minimum that an educated person should know about the ubiquitous computer.

The phrase "computer literacy" here means that people acquire enough understanding of common technical terms so they are not intimidated by computer jargon. This chapter should help you acquire some technological basics to make better use of your home or workplace computer.

Personal computer systems are usually made up of three components: the display screen or monitor, the keyboard, and the metal box that contains the specialized circuits that do the work—that is, process our words, balance our accounts, store our data, and revise our business forecasts. Other equipment, called *peripherals,* may include a printer, a disk drive, and a control device called a mouse. Figure 3-1 illustrates the components of an average personal computer system.

If you, like many others, have been faking it—pretending to know all about computers—this is the chapter for you. Read it in secret and catch up. Although you can drive a car to the supermarket without

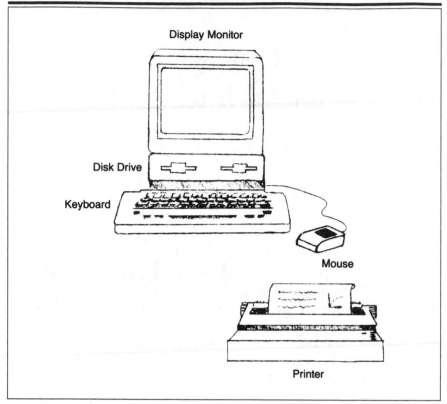

FIGURE 3–1 *Typical Symbol Manipulation System*

knowing the first law of thermodynamics, there are fundamentals with which you have to be familiar to make an intelligent buying choice in the car market: fuel efficiency, safety features, structural integrity, and engine and transmission types, to name a few.

When it comes to buying a computer, a similar knowledge of fundamentals is necessary to avoid buying more or less capability than you need. It is difficult to read the computer ads. If you don't know what RAM, ROM, kilobyte, disk drive, and memory mean, you feel lost. To say nothing of bits, bytes, ASCII, CP/M®, MS-DOS®, and OS/2®. And it is getting worse. This is how Motorola advertises a new system:

The 88000 runs at a blistering 14–17 MIPS, 7 million floating-point operations per second, and an incredible 50 MIPS in parallel processing applications (using just four 88000 chips set on our HYPERcard).

If that doesn't sell you, try this one from Toshiba:

Our T3200 has the advantage of a 12MHz 80286 microprocessor, an EGA display system, a 40MB hard disk, and 1MB of RAM expandable to 4MB. Also, its two IBM-compatible internal expansion slots let you connect your PC to mainframes, LANs, and more.

More I couldn't take, but let's not let the jargon jungle hold us up. With a sharp machete and an experienced guide we'll get through.

In 1833 English inventor Charles Babbage conceived what he called the "analytical engine," which would not only perform arithmetical calculations but also make decisions based on the results obtained. By combining arithmetical and logical functions, Babbage in effect formulated the first programmable computer.

Babbage's "engine" was totally mechanical, running on gears and levers, but it did incorporate a storage unit and an arithmetic unit, and it was designed to compare quantities and execute different instructions based on what we would now call a program. Also, the results of any operation could be automatically fed back to control the next step in a complicated sequence of operations. It even used punched cards as input.

At least in theory it could do all of these things. The sad fact is that neither of the two machines Babbage built actually worked. To cynics, this makes Babbage the father of the modern computer in more ways than one. It was, however, a brilliant concept as it sought to do with gears and levers what computers of today do with microcircuits.

Consider one more piece of history before you take the cover off your computer. About the same time that Charles Babbage was working to perfect his machine, a contemporary of his named George Boole developed symbolic logic, reducing all logical relationships to simple expressions—AND, OR, and NOT. Boole showed how the types of statements employed in deductive logic could be represented by symbols, and these symbols could be manipulated according to fixed rules to yield logical conclusions. Boolean algebra, as it came to be called, permitted the automation of logic by operating with combinations of

0s and 1s. Binary switching—the basis of electronic computers—was thus made possible.

It is time now to look at the magical world of 1s and 0s.

Binary Thinking and the Wonderful World of 1s and 0s

You and I communicate our thoughts by combinations of 26 letters and 10 digits (and sometimes pictures). The computer, however, is limited to on/off circuits, so how does it carry out its sometimes complex tasks? Machines don't think, but they come close to thinking and they do it just by using 1s and 0s. Is it absolutely necessary to know a little about the binary system in order to use a computer intelligently? No, you can go on thinking the whole thing works by magic, but if you want to speak computerese like a native, you have to learn about bits, bytes, and words.

The term *bit* is a contraction of *binary digit* and it is the smallest unit of the binary system. A bit is either a 1 or a 0, depending on whether it is *on* or *off*. Imagine, if you will, a Christmas tree light bulb that can be either on or off.

When we string a group of these bits (lights) together it is called a *byte* and it is the basic unit of information in the computer. A byte usually contains 8 bits. Because each bit has 2 states and there are 8 bits in a byte, the total combinations available then are 2^8, or 256.

Now how does a computer use these bits and bytes to store and manipulate information? We commonly communicate with the computer by pressing keys on the typewriter-like keyboard, and the computer communicates with us by displaying characters such as letters, digits, and punctuation marks on a television-like screen. To permit the computer to process data in character form, we need some way of representing characters in terms of bits. This *character code* is similar to the Morse code used in transmitting telegraph messages.

Most computers use a code system called the American Standard Code for Information Interchange (abbreviated ASCII and pronounced *as'key*). The ASCII code uses 7 bits to represent each character and can represent 128 characters in all. These include the upper-

and lowercase alphabets, the digits, the punctuation marks, a few mathematical and other signs, and a set of control characters used to facilitate data transmission. The 7-bit ASCII characters fit nicely into an 8-bit byte. ASCII characters are therefore normally stored 1 per byte. (The eighth bit is used as part of the internal machine code.) Using our string of Christmas tree lights (1 byte), the letter M would be as shown in Figure 3-2. The message "Merry Christmas" coded in ASCII is shown in Figure 3-3.

Computers can move around more than one byte at a time. These larger units are called *words* and usually contain 2 bytes or 4 bytes. Word lengths are most often expressed in bits, so the equivalent word length is 16 bits or 32 bits. When you are told that a computer uses

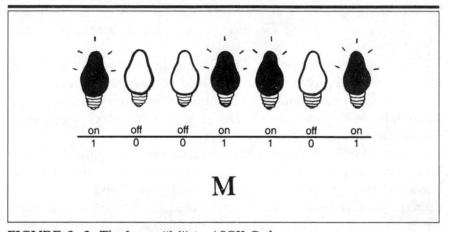

FIGURE 3-2 *The Letter "M" in ASCII Code*

| 1001101 | 1000101 | 1010010 | 1010010 | 1011001 |

M E R R Y

| 1000011 | 1001000 | 1010010 | 1001001 | 1010011 | 1010100 | 1001101 | 1000001 | 1010011 |

C H R I S T M A S

FIGURE 3-3 *The Message "Merry Christmas" in ASCII Code*

a 32-bit microprocessor, it means that the system is capable of moving around data in hunks 4 bytes in length. The message is that a 32-bit microprocessor is more efficient and faster than a 16-bit microprocessor.

The binary system seems cumbersome in comparison to our own base-ten or decimal numbering system. For instance, it takes only four digits to represent 1989 in decimal numbers, whereas it takes four bytes of eight bits each to represent the same number in binary. Problems in addition, subtraction, multiplication, or division also seem very cumbersome in binary when compared to our base-ten system. Why use it then? The answer is that the machine demands that entries be made as simply as possible. In fact the machine requires an on/off, 1/0 system. That's the bad news. The good news is that the machine handles these apparently cumbersome units so fast and so tirelessly that, like Annie in *Annie Get Your Gun,* it can do anything we can do better (almost).

The English mathematician Alan Turing demonstrated in 1936 that any mathematical problem could be solved by a machine if it could be expressed in a finite number of manipulations. Today, using supercomputers, math problems of mind-boggling complexity are being solved, and the base system is still 1s and 0s.

If instead of 1s and 0s we think in terms of true or false, the same binary system applies as a logic tool. I mentioned earlier that George Boole conceived a system of logic that took "true" and "false" away from the philosophers and gave it to the mathematicians. Boolean logic uses a device known as a truth table, which lists all possible combinations of true and false values that can accrue in the interplay of two or more statements. In Boolean algebra, statements can be weighed and their consequences evaluated with the help of the three logical functions, AND, OR, and NOT.

A computer can handle logic problems as easily, and in the same way, as it handles mathematical calculations. As an example, take the case of the statement "Either A or B is true." In this case, the problem can be handled by addition of binary symbols. The binary digit 1 is equal to true; 0 equals false. If both A and B are false, the statement expressed in binary digits would look like this:

$$0 + 0 = 0$$

Expressed in words, this statement is false. If A is true and B is false, or vice versa, the Boolean algebra looks like this:

$$1 + 0 = 1, \text{ or } 0 + 1 = 1$$

In words, this statement is true. Other logic problems require multiplication or division, but in any case, the types of statements employed in deductive logic can be represented by mathematical symbols, and the symbols can be manipulated according to fixed rules to yield appropriate conclusions.

Babbage's analytical engine attempted to carry out all of the required mathematical and logic functions by means of gears and levers. Modern computers use a microchip, but just what is a chip, micro or otherwise?

Because they are made with chips or small slices shaved from a crystal of silicon, microelectronic circuits are known generically as *chips*. The word *semiconductor* is used to describe substances such as silicon or germanium. These substances have electrical conductivity capabilities intermediate between that of an insulator and a conductor. That is, an area of the chip can be chemically treated to readily conduct electricity, whereas a different area of the chip can be made to prevent the flow of current.

In general, the chips in a computer are composed of transistors, which function as switches. Electrical current goes from one end of the device to the other, crossing through a gate that sometimes lets the current pass and sometimes does not, providing the on/off signals needed for a binary system.

The best way of making chips faster has been to make them smaller, giving the current less distance to travel. Thus, developers have created the very small "microchip." This shrinking has allowed computers to become faster as well as smaller and cheaper.

The time has now come to see what's inside a personal computer.

ANATOMY OF A PERSONAL COMPUTER

Computers have been called information processing machines. But computers cannot process information in the abstract; they can only manipulate the symbols of information, such as letters or numbers. A computer might better be described, then, as a symbol-manipulating machine. It is a mistake to think of a computer as a thinking machine; it is not. The computer user has to think. Yes, the computer can help by doing all the drudgery, like adding long columns of numbers or putting the names on a long list in alphabetical order. But those actions require no thinking, they involve merely juggling bits of information.

The computer's capability to juggle and manipulate depends on four basic functions: input, processing, memory, and output as shown in Figure 3-4. External to the computer are storage systems such as hard or floppy disks and printers. My emphasis here is on the internal functions, each discussed separately. Because a computer will manipulate meaningless symbols just as efficiently as meaningful ones, "garbage in, garbage out" became an often used cliche in the computer field. But how do we get information, meaningful or not, into the machine?

Input

Computers accept information from a variety of sources: typing on a keyboard, drawing with a lightpen, moving a mouse, pressure of a finger on a display screen, output of a sensor device, scanners like those used at the supermarket checkout stand, or even words spoken into a microphone.

The three most commonly used input devices are

1. the alphanumeric or typewriter-like keyboard.

2. the mouse (not a rodent, but rather a palm-sized, button-operated device used to move the cursor on the screen).

3. continuous signal sensors or telephone line from another computer.

The Keyboard

A computer keyboard looks like part of a typewriter but bears little resemblance to it except superficially. The letter C on a typewriter is that and nothing more. The same letter C on a computer keyboard can be used for a number of functions in addition to typing the third letter of the alphabet. This is possible because pressing a key on the computer's keyboard initiates an electronic signal that can be assigned a wide variety of meanings.

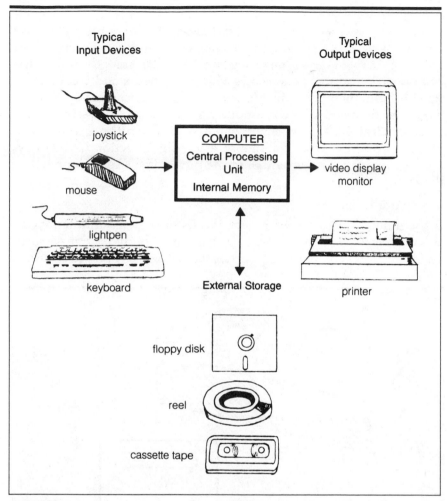

FIGURE 3–4 *A Computer System, Including Peripherals*

When the computer detects a keystroke, it uses a look-up table to determine the code assigned to that key. As explained earlier, the code is a binary number assigned in accordance with an accepted coding system called ASCII. For instance, if I press the A key on the keyboard, the computer registers 1000001. A different code can be assigned to a key when it is pressed in conjunction with other keys, and this ability is what gives the computer keyboard so much more flexibility than a typewriter keyboard.

The Mouse

Figure 3-5 shows a typical mouse. It usually contains a rolling ball and one or more buttons used to execute commands or select text or graphics. As the mouse is moved around on a flat, smooth surface, the ball underneath moves and sends what are called x and y coordinate signals to the computer. Moving the mouse causes the cursor (flashing indicator) to travel in a corresponding manner on the screen. The cursor indicates where an action will take place on the screen, and it can be moved by either the mouse or by the direction (arrow) keys on the keyboard. The advantage of the mouse is the speed with which the cursor can be maneuvered.

Continuous Signal

If the input to the computer is by means of some continuous signal (a temperature or pressure sensor), the signal must be translated to something the computer can understand—digits. The translator is

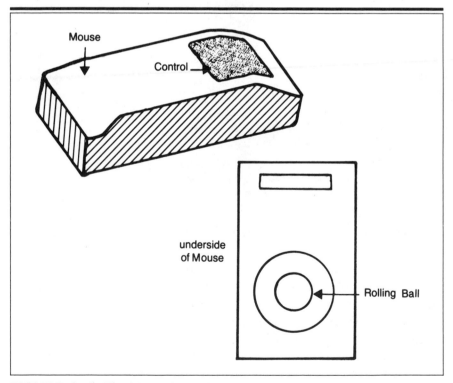

FIGURE 3–5 *The Mouse Input Device*

called an analog-to-digital converter. *Analog* means analogous to or representative of some reality, whereas *digital* means having to do with numbers. The first counting tools humans ever used were their fingers, and the word *digit* derives from words for finger (or toe) and for a numerical integer. Digital clocks or digital speedometers display information to us by means of numbers or digits.

Analog information, on the other hand, is continuous. If I plot temperature versus time, for example, the result will be a continuous curve (see Figure 3–6). Because, as we have seen, computers employ two-state or binary logic, all input must be translated to discrete 1s and 0s or on/off terms. Remember that in essence, the computer is just a bunch of on/off switches arranged in ingenious order. The A-to-D converter does its job by taking periodic samples of the incoming continuous signal and translating the data to corresponding digital data.

Once translated to binary pulses, information enters the computer through electronic gateways called *ports*. A port may be either serial or parallel. If it is serial, the binary pulses enter the computer system in single file, and if it is parallel, the pulses are lined up abreast at the port to form a set of pulses of whatever length the computer is designed to handle. As is rather obvious, parallel input (or output) is much faster.

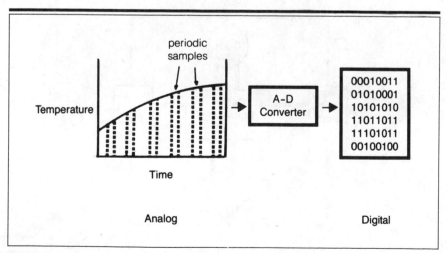

FIGURE 3–6 *Analog-to-Digital Converter*

Information Handling

The internal pathways for information in the computer are called *buses*. A bus is nothing more than a connecting circuit. Instead of running wires between each component in the system, all the parts are connected to a common set of wires—the bus. Control signals travel on the control bus; data travels on the data bus to a destination specified by the address carried on the address bus.

Most computers have a bus that is 8 or 16 bits wide, because the microprocessor they are built around can process only 8 or 16 bits of information at a time. Newer microprocessors, like the Intel 80386, can process 32 bits of information at once, so a 32-bit bus is needed to take full advantage of the processor power.

Central Processing Unit and Memory

Now that you know how the signal or data reaches the computer, it's time to meet CPU, RAM, and ROM (see Figure 3–7). The *central processing unit (CPU)* is the administrator of the machine. It controls all of the functions of the computer and makes all the calculations. The CPU is heavily dependent on information and instructions stored

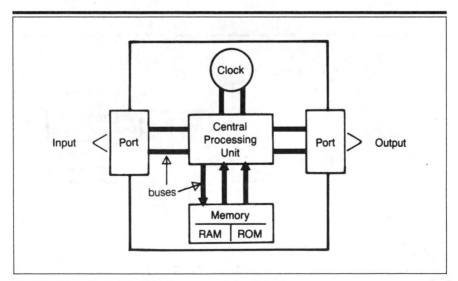

FIGURE 3–7 *Inside the Computer*

in two types of computer memory: ROM, read-only memory; and RAM, random access memory. CPU, ROM, and RAM are the team that runs this symbol-manipulating machine, so you should know a little more about each.

The CPU coordinates all the computer's activities, for example, interpreting commands entered from the keyboard, retrieving data from a memory bank, or sending data to a printer. The CPU performs the arithmetic calculations such as addition and subtraction using the binary system of mathematics. It also performs the logical operations using equal to, greater than, less than comparisons. For example, it can be programmed to determine average household incomes in a town and print out a list that ranks all households in order by income level or any of a number of like information-juggling tasks.

CPU is the CEO of this firm, but the two staff assistants, ROM and RAM, are necessary to make the computer operate efficiently. The two assistants differ in personality in important ways. Old ROM is important to have around because he knows a lot of useful operating information but he is fixed in his ways. RAM is the really smart assistant. He is more flexible and more able to handle new situations, but if you are not careful he will discard your information.

ROM is the computer's permanent internal memory. We can use information stored there, but we can't add anything to it. Here "read" means to retrieve data from a memory, and "write" means put something into memory. That is how the name read-only memory was created for this component. ROM holds essential operating information such as startup instructions.

RAM is the real indicator of a computer's I.Q. When computers are being compared, the amount of RAM available is a key feature. A 640K RAM computer is smarter than one with 64K. The K means kilobyte (K is an abbreviation for the kilo, which is Greek for a thousand). A thousand bytes is the equivalent of about half a page of double-spaced text. A megabyte is a thousand K (or 1 million bytes). A computer with 2MB (megabytes) of RAM has several times the intellectual capability of one with 640K and can accommodate smarter or more complex software programs than the one with only 640K. Generally speaking, the more complex and lengthy the program, the more space in RAM it requires. Programs sold at the software store always specify the amount of RAM required.

Another important difference between ROM and RAM that many of us learn the hard way: RAM is the temporary memory bank, the repository of programs and data that are used by the CPU. *Temporary* is the key word here. Information stored in RAM is wiped out from the computer when the power is turned off. It is lost forever—a tragedy to be avoided. Information in RAM must be saved by storing it on a floppy or hard disk. When you save data, you are moving it from one storage medium to another (RAM to disk), just as when you take the paper on which you have been typing and place it in a file cabinet. The disk also has a memory capacity expressed in kilo- or megabytes. A floppy disk, depending on its size, can hold from 60 to 600 pages of typewritten material. A *hard disk* resembles a much bigger filing cabinet and can store 20, 40, or hundreds of megabytes of information. One 40MB hard disk equals 114 floppy disks that each store 360 kilobytes. As you can see, your computer may have various types of memory: 640K RAM, 64K ROM internally, 360K floppy, and 20MB hard externally, for instance. Now you know the difference and you are in a better position to understand the computer ads.

In addition to entry ports, buses, CPU, and memory systems, a computer has a clock component that regulates and synchronizes the activities of the computer by means of regular electronic pulses sent through the control bus. The clock controls the speed with which the computer operates and keeps everything in order. If too many actions try to happen at the same time, confusion results and data is lost. Clock speed is important because the faster the clock runs, the faster the computer can process data. Clock speed rates are expressed in units called *megahertz (MHz)*—or millions of cycles per second. When a computer is advertised as having a 12MHz microprocessor we know it is twice as fast as one having a 6MHz microprocessor. Let's assume that a computer has done its symbol manipulation function and now we want to get the results.

Output

The most common output devices are video display monitors and printers. Other output devices include modems used to transmit information over telephone lines and plotters used to develop graphic representations.

Video Display Monitor

The video display monitor, often called CRT for cathode ray tube, gives you instant feedback while you are working. What you see on this screen is called *soft copy*. Besides displaying text, monitor screens can also display graphics when the computer is configured to handle them.

Printers

There are three different types of printers:

- Dot-matrix, in which pins strike the printer ribbon to create letters from series of dots

- Daisy-wheel or letter quality, which use fully formed letters on wheels or thimble-shaped heads to impact the printer ribbon a whole character at a time

- Laser, which is a nonimpact system producing an entire page at a time

Printers have improved in the last few years to the point that even the least expensive dot-matrix type provides good quality.

Dot-matrix printers form characters using dots; and the spacing of the dots, that is, the closeness of the dots, determines the quality of the printed material. Speed of operation is what determines the cost of dot-matrix printers. Better printers can deliver over 200 characters per second (CPS) at draft speeds (lower quality) and 80 CPS at near-letter-quality speed (more than two double-spaced pages per minute).

Daisy-wheel printers achieve typewriter quality because they use fully formed characters and print a single character at a time. They employ a wheel-like component to move the selected character to the proper striking position. Daisy-wheel printers are generally slower than dot-matrix ones but produce a higher-quality output.

Laser printers are the most expensive of the three printer types. They form characters using dots, but because they use many more dots, the sharpness of the finished quality is better than the dot-matrix printer. Laser printers are based on an electrophotographic technology similar to an office copier. Because they print a page at a time, lasers are by far the fastest of the printers available.

There exists a speed gap between computers and printers, of whatever type, and a device called a *buffer* is therefore required to bridge this gap. The buffer receives data from the computer as fast as it is produced, stores it, and feeds it to the slower printer as needed.

The foregoing anatomy of a computer and its peripherals (all the units outside the main "black box") provides an overview of the system. Those parts of the system that you can hit with a hammer (not advised) are called *hardware*; those program instructions that you can only curse at are *software*.

So far I have been concentrating on what the hardware in the machine can and cannot do. It is now time to turn to the language of computers, the instructional programs called software.

COMPUTER LANGUAGES

Software aficionados tell us to ignore the hardware of the computer. The true computer, they tell us, is not its machinery, any more than music is defined by its instruments. As a musical score is to a symphony, software is to the computer. Software freaks point out that the computer is the first machine invented by humankind that has no specific purpose other than to follow instructions.

Three kinds of software are required to get the most out of our modern computers:

- An operating system to control the computer

- A programming language used by programmers to write programs

- Applications programs of the type we can buy at the software store to perform a specific task for us (such as word processing or accounting)

Operating Systems

An *operating system* coordinates activity between you and the microprocessor and between the microprocessor and other parts of the computer system. The operating system controls your computer, usually without your involvement or awareness. Operating systems are built into the computer at the factory and they come in several standard types.

The best known and most used operating system is MS-DOS—which is an abbreviation for Microsoft® Disk Operating System. Most IBM® machines, except the newest ones, use a version of MS-DOS called PC DOS®. MS-DOS and PC DOS are so similar that software written for one will almost always work with the other. The so-called IBM clones—computers similar to IBMs—use MS-DOS.

Apple® and Macintosh® operating systems are both proprietary and unlike any other systems. The Apple system is called Apple DOS™, whereas the hard disk version is called Pro DOS. The Macintosh system is designed to be supereasy to use. It uses small representative figures called *icons* instead of commands.

CP/M (Control Program for Microcomputers) was the forerunner of operating systems. At one time it was the most widely used microcomputer operating system for 8-bit computers. The current dominance of IBM and IBM-compatible equipment has moved CP/M to the backwaters of the computer world. There are still plenty of CP/M machines in operation, but developers aren't writing much new software for them.

OS/2 is the latest operating system. It was developed by IBM to run on their newest PS/2® (Personal System/2®). OS/2 works only on machines using the newer 80286 or 80386 microprocessors. Furthermore, the machines must be equipped with at least 2 megabytes of memory—in other words, very advanced machines. OS/2 takes advantage of the 80286/386 power and speed to run very large (and complex) programs, and it allows several programs to run simultaneously. You can't do either with DOS.

A third system, the UNIX® operating system, is not tied to any specific microprocessor. Often the UNIX system is preferred when many computers of different sizes are tied into one network.

What's important to understand about all this is that operating systems are specific to the hardware and software with which they are used. You cannot simply insert MS-DOS software in an Apple and use the program, nor can Apple or Macintosh software be used directly with a non-Apple PC.

Programming Languages

Computers use a hierarchy of programming languages. At the low end are the languages the machines use, and at the higher level are the languages the programmers use to communicate with the machine.

Machine language consists of the operation codes built into the machine (in 0s and 1s, as explained earlier). Machine language programs are cumbersome and lengthy and are not well suited for easy modification.

The next step up in the language hierarchy is *assembly language*, which can be thought of as "machine language without tears." Here the programmer can use commands in the form of meaningful abbreviations called *mnemonics*. For example, assembly language uses MOV to mean move, CLR for clear, or RET for return. Here again, remember that computers understand only binary instructions, so although the programmer uses assembly language, the computer uses a decoder and translates all instructions back into machine language. Assembly language is easier to understand than machine language, but it is still too awkward for everyday programming.

Assembly language is oriented toward the machine rather than toward the problem being solved. Higher level languages are oriented toward the user's needs and come in many different flavors.

Fortran, for instance, was the "Latin" of computer languages used by engineers who were primarily interested in processing formulas. This met the needs of scientists and engineers but was not suited to business needs. So industry users and computer manufacturers got together and produced COBOL—Common Business Oriented Language.

Because a language was needed for students and home computer users, BASIC was invented. BASIC stands for Beginners All-purpose Symbolic Instruction Code and is an easy-to-learn programming language that is available on almost all microcomputers.

As more people bought home computers, the industry became what was known as "Hoosier friendly," and the simplified languages took on such names as Pascal, SMALLTALK, and THINKLAB. Some suggested that we were on our way to an all-purpose simplistic language called GRUNT, but I think they were kidding.

Application Programs

Application programs are the kind most computer users use directly. They are available at software stores, and they are programs designed for some specific use. Common applications include word processing programs such as WordStar® or spreadsheet programs such as Lotus

1-2-3, ® for example. Applications are written to run with a specific operating system. A program written to work with one operating system generally does not work with another. To match the application with your operating system, check the box the software comes in because most new programs are written in several versions. Buy the version designed to run with your operating system.

SPEECH RECOGNITION

Computer-based speech recognition is already well established. Systems that recognize single words or simple phrases from a limited vocabulary are being used in factories for controlling machinery, entering data, inspecting parts, or taking inventory. In some hospitals doctors and nurses can keep their hands free for working with critically ill patients by wearing microphones so that they can describe their actions to a computer that logs the information and keeps the necessary records.

A more advanced speech recognition system called SPHINX has been developed with a vocabulary of 1000 words. Using this system, Pentagon planners can speedily find specific information stored in a database by simply asking the computer to make the search.

SPHINX understands continuous human speech with at least 97 percent accuracy and does not have to be taught to recognize each user's voice. This system will probably see its first commercial uses in voice-controlled systems in military aircraft and in computer equipment used by engineers and industrial designers. Researchers are currently working to increase SPHINX's vocabulary to more than 20,000 words. Eventually, it may be possible to use the system to help translate spoken language, a long-sought goal for computer engineers.

Completely speech-controlled home computers are the goal of advanced planners, but the technology is still a long way from that objective today.

GRAPHICS

There are two kinds of computer graphics programs:

- business graphic programs that you can use to create charts and graphs to illustrate your business reports or create overhead transparencies

- interactive graphics programs that are used by designers, artists, or engineers to design on the screen. They are interactive because the designer can draw right on the screen—seeing the results of changes or manipulations in real time

You must be subjected to another attack of jargon here. Graphics standards are most often described by letter triplets such as EGA, VGA, MGA, and so on. You will remember the Toshiba advertisement back at the beginning of this sojourn in computerland. One of the features of the system they were touting was "an EGA display system."

There is a theory that computer manufacturers purposely put a lot of jargon in their ads to screen out the riffraff (like you and me) and limit inquiries to the knowledgeable. Let's fool them and cut our way through the specialized vocabulary.

Graphics are either monochromatic—ordinary black and white— or color. For color, IBM set the standard back in 1984 with the EGA (Enhanced Graphics Adapter). In 1987 IBM introduced VGA (Virtual Graphics Array) along with its PS/2 computers, and VGA became the new standard for most non-Apple computers.

VGA is considered superior because its resolution and sharpness of image are so much better. VGA supports 640 picture elements (pels, pixels, or dots) horizontally by 480 pels vertically. The more elements used, the better the picture.

IBM PS/2 Models 25 and 30 have still another system called MGA (Multicolor Graphics Adapter), but VGA is expected to remain the standard for the near future. The big advantage of VGA is the wide range of colors you can place on the screen. VGA can display up to 256 colors, but the number of colors is not as important as the combination of colors and resolution: color always comes at the expense of resolution. At full resolution, VGA can display 16 colors at once. In its monochrome mode, VGA can display about eight dozen shades of gray. What can you do with all this capability? You can devise pretty impressive illustrations for documents, for one thing. Graphics can be a great communication tool, limited only by your imagination.

CAD/CAM

Good grief, not more jargon! I promise to add only what is necessary to understand computer aided design (CAD) and computer aided manufacturing (CAM). Both CAD and CAM are interactive graphics programs, permitting the designer or engineer to make changes and see the results displayed on the screen.

Engineers, architects, and designers use CAD programs to speed up the conceptualization of designs and their conversion to working drawings. They can, in effect, create scale models on the screen, increase or decrease their scale, rotate them in two or three dimensions, and change them at will.

CAM programs are used to produce instructions for the computers that control machinery in a manufacturing process. That is all you really have to know about CAD and CAM unless you are going to work with them on a regular basis, in which case you will have to know about workstations.

WORKSTATIONS

Larger organizations often connect a number of computers to a central computer so that the central CPU is shared by all the workstations. Workstations are either dumb or smart. *Dumb terminals* don't have their own microprocessors, and when disconnected from the network, they are nothing but a piece of furniture that looks like a computer system but isn't. *Smart workstations* are microcomputers with their own memory and storage systems linked both to a central CPU and to other workstations. When disconnected from the link, smart terminals can run their own software.

Many workstations are hard to distinguish from personal computers. Workstations have a screen, keyboard, central processing unit (at least the smart ones do), and maybe a mouse. What distinguishes a workstation from a personal computer is not the exterior appearance but what's inside the box. Specifically, workstation features include increased power, the ability to run several programs simultaneously, advanced networking abilities, and enhanced graphics.

It is mostly the networking capability that gives a workstation its superior power. The network gives rapid access to other users (and the material in their systems), to massive data storage, and, sometimes, to mainframe or even supercomputers. Consequently, workstations users have at their fingertips computing power exceeding that of a personal computer by several orders of magnitude. Workstation speed is measured in MIPS (millions of instructions per second). Memory capability is measured in megabytes, and display quality is measured in megapixels. Tenfold leaps in capability for workstations in all three areas are predicted by industry leaders by 1995. Speaking of enhanced power, let's turn now to the superfast and superpowerful.

SUPERCOMPUTERS

What distinguishes supercomputers from the humble home computer is speed and power. Costs might also deter you from considering the purchase of a supercomputer—they run between $5 million and $25 million each. Supercomputers are giant number crunchers capable of solving mind-boggling problems heretofore considered too complex even for computers: accurate long-range weather forecasting which must take into account thousands of detailed, continually changing phenomena, or designing the hypersonic aircraft of the future, to name but two uses.

FLOPS, gigaFLOPS, and teraFLOPS

Speed is the major feature of supercomputers. In the early days of the big machines, speed was measured in thousands of FLOPS, an acronym for floating-point operations per second, in which the decimal point is moved in very large and very small numbers. The speed of the largest machines today is measured in gigaFLOPS, or billions of operations per second. The more advanced machines of tomorrow will be measured in teraFLOPS, or trillions of operations per second. A supercomputer in the teraFLOP class will have the power and speed of 10 million personal computers. They achieve this impressive capability with a system called parallel processing.

Parallel Processing

The ordinary computers I have been describing in this chapter have one microprocessor—usually on a single microchip, or integrated circuit—that can work on a problem only one step at a time. Parallel computers divide the problem into smaller pieces and assign each piece to one of many processors working simultaneously.

If, for example, you tried to manufacture an automobile by yourself, doing one task at a time, it would take forever. But if the various parts were manufactured separately and then brought together on an assembly line, the task could be finished much sooner.

The number of parallel processors used in a supercomputer seems to be the dominant factor. The CRAY-2®, one of the best known and most widely used supercomputers, uses 4 parallel processors. The CRAY-3® has 16, and the CRAY-4 will have 64.

Fastest of all the supers is the one developed by scientists at the Sandia National Laboratory in New Mexico. Their supercomputer can solve complex problems 1000 times faster than a standard computer, a speed much faster than scientists had believed possible. The Sandia machine consists of 1024 processors linked in parallel.

IBM has a super-supercomputer in the works called the TF-1 that will exceed in speed and power any computer ever built. Designers say that when completed, the TF-1 will be capable of speeds 2000 times as fast as today's supercomputers.

The trick is not just to add more and more parallel processors, however. The real problem is writing a program that can break a complex problem into parts that can be worked on simultaneously. The program must identify all the steps that must be accomplished, assign the relevant data and programming to each processor, coordinate communication between the processors, and then assemble the final results—in short, supersoftware.

Where is this discussion leading? Many scientists believe that the country that leads the world in supercomputers will hold the key to technological and economic development in the 1990s and beyond. Another important issue is the role of supercomputers in the development of artificial intelligence.

WHAT IS ARTIFICIAL INTELLIGENCE?

Chapter 2 described how molecular biologists explained how four nucleotides form a language that tells a cell how to arrange twenty amino acids into a complex system we call life. *Artificial intelligence (AI)* is about how symbols representing things in the world can be integrated into machines that mimic the language of the mind, the process we call thinking.

AI covers a broad base of technologies that include such fields as speech recognition and synthesis, natural language processing, and pattern recognition. Most of the attention so far has been on "expert systems" programs—software that represents the knowledge of human experts in various fields. Expert systems are based on "if/then" rules, which in conjunction with facts about a particular subject can logically progress through a given problem to arrive at an appropriate solution. Expert systems are meant to mimic the knowledge, procedures, and decision-making powers of real people. We have seen in this chapter so far how computers have automated data processing. AI proposes to automate decision making.

The core concept in artificial intelligence is the use of *heuristics*, sometimes known as the art of good guessing. Heuristics allow computers to deal with situations that cannot be reduced to mathematical formulas and may involve many exceptions. Consider, if you will, the mental processes of a chess player. The expert player ignores the almost-infinite number of moves and concentrates only on those relevant to a particular situation. Similarly, expert system programs do not consider every possible variation in a problem, but only those likely to lead to a solution.

The method used to put the "expert" in expert systems in building an AI program is called *knowledge engineering* and it is a time-consuming, expensive, and complicated process. Knowledge engineers meet with experts in a particular field, such as auto mechanics or medicine, and pick their brains—particularly the problem-solving methods they use. From these brain-picking sessions, the knowledge engineers come up with a set of rules applicable to a particular field. The rules become the basis for a problem-solving computer program.

Expert systems apply expert decision-making rules and a base of knowledge to solve problems like a trained person. In their simplest form they are decision trees that say, "If X, try W. If Y, try Z." More complicated systems will use advanced techniques to search through thousands of possible answers to solve a problem in an efficient way.

How well are they doing? Let's consider a few examples of AI in the field.

A medical diagnostic program developed at the University of Pittsburgh was tested using cases from the *New England Journal of Med-*

icine. The AI program was then compared to that of a group of physicians. Over a wide range of diagnoses, the AI system was found to be more accurate than was the average physician.

Both Ford and General Motors are now using expert systems. Ford mechanics across the country can tie into a computer network AI system designed to mimic the thinking that one of their engine experts uses to solve hard-to-diagnose engine problems. GM uses an AI system to duplicate the thinking process of one of their now-retired machine-tool troubleshooters. In each case, the AI system allows less-experienced employees to make decisions like an old pro.

At American Express, an AI system assists less-experienced financial loan officers to evaluate credit requests. The system clones the expertise of a number of specialists in one expert system.

AI has not been without its critics. Some claim the whole idea has been overhyped. In its early days, AI was the black hole of the computer world; you were asked to believe in it even though you could not see it. Princeton's distinguished scientist/writer, Freeman Dyson, said that artificial intelligence was certainly artificial all right, but there was doubt about its intelligence. Other critics called it artificial insanity.

The more recent successes of AI have won over some doubters, at least in the area of expert systems. Success is measured in the marketplace. The number of AI systems in use in the United States is now increasing by 50 percent annually.

NEURAL NETWORKS

Neural networks are computer models of how the brain might work. Untangling the brain's complex circuitry in order to decipher how animals and humans store and recall information is a difficult biological task. In the last few decades, scientists have made significant advances in determining chemical and physical changes in individual neurons thought to be involved in memory. By observing behavior, the researchers have classified forms of memory and linked them to different regions of the brain.

The next step, understanding not just how one neuron but the hundreds of thousands of nerve cells interact within the brain, has brought about a remarkable joint effort among neurobiologists, psychologists, computer scientists, physicists, and philosophers.

Proponents of neural network efforts reason that if scientists cannot directly observe and dissect memory and learning processes in biological brains, perhaps silicon ones, able to handle highly complex webs of interconnected processing elements, may help fill in the gaps.

Their ultimate aim is to explain how the brain uses electrical and chemical signals to represent and process information. Science is now in position to serve as a matchmaker between the computer hardware of neural networks and the three pounds of "wetware" encased in the human skull.

At the philosophical heart of network modeling lies the idea that the mind emerges from the brain's behavior. It makes sense, then, to imitate in computer setups the structure and biological wiring of the brain to reproduce mental abilities.

How well are they doing? The science, it must be said, is at a primitive stage. Perhaps the most successful of the neural-net models to date is the one developed at the University of California at Irvine by neurobiologist Gary Lynch, computer scientist Richard Granger, and their colleagues. Together they have built a simulation of one part of a rat's olfactory cortex. The simulation model illustrates how the brain identifies and remembers smells.

Despite its relative simplicity, researchers believe that biologically based computer simulations of this region of the brain may lead to a deeper understanding of the much more complex neocortex. The neocortex accounts for 80 percent of the human brain and is thought to play a role in everything from language processing to spatial tasks. The neocortex is thought to have evolved from and retained many of the features of the more primitive olfactory region.

Neural networks may in time lead scientists to understanding the circuit diagram in our brains. In the process of this research, the researchers may even develop machines that really do think.

CONCLUSION

In conclusion, the role of the computer in human affairs has yet to be fully imagined. Because these machines are going to play an increasing role in our lives, you will be in a better position to take advantage of them now that you have learned a little about what makes them tick (or in this case, hum).

KEY CONCEPTS

▶ Computers are collections of on/off switches wired together in circuits varying in complexity depending on the capacity and capability required.

▶ Because they are made with chips of silicon—a semiconductive material—microelectronic circuits are known generically as chips or microchips. In general, chips are composed of transistors, which function as switches.

▶ Computers do all their work by using 1s and 0s (bits) in combinations (bytes and words) that form a code. This binary system can represent all of the letters of the alphabet, all the numbers, and all the characters that we need for the many tasks we ask the computer to do for us.

▶ The CPU, or central processing unit, coordinates all of the computer's activities and is assisted in this task by ROM, or read-only memory, and RAM, or random access memory. ROM is like a phonograph record in that information can be taken from it at any time, but no new information can be added to it. RAM is like a cassette tape in that information can be added, altered, or retrieved at any time.

▶ Software consists of step-by-step programs that direct the machines (hardware) that make up the computer system. Three kinds of software are needed: operating systems that the computer understands, programming languages used by experts to write programs for us, and application programs of the type we buy to perform a specific task.

▶ The predictions of 10 years ago that research into artificial intelligence (AI) would produce computers with the intelligence of human beings have not come to pass. Many scientists still envision a bright future for AI eventually, but goals for the next decade have become considerably more modest than they were formerly.

4

Environmental Penalties
of
HIGH
TECHNOLOGY

WHAT'S HAPPENING to our world? Unless you have been living in Timbuktu for the last few years, you know that there is growing scientific concern about our global environment. The Chicken Littles have warned us for decades about environmental abuse. They have been predicting imminent calamity for so long that we tend to discount them. But could the Chicken Littles be right this time? Have they been right all along?

This chapter covers four major environmental penalties of high technology:

- greenhouse effects
- ozone depletion
- acid rain
- waste disposal

There has been much discussion in the popular press about each of these threats to the Earth's environment but not a lot of agreement. We know these problems are largely global in nature and that the nations of the world have yet to develop a policy featuring overall coordination.

Environmentalism is changing from a local, usually noble cause involving birdwatchers and Sierra Club members to an issue of serious international concern capable of drastically affecting the life style of each of us.

This chapter is not all doom and gloom, because although misuse of technology may have gotten us into trouble, technological innovation can help extricate us. Reversing the damage (where it can be reversed) will be expensive and may change world economic patterns. If what a large body of scientists are now telling us is factual, we have no choice but to undertake global remedial actions.

At the moment there is no scientific consensus about what is really happening and what proportion of the damage comes from industrial emissions, automobiles, global deforestation and changing land use, and use of chemical compounds in the home and on the farm.

Before any action on a significant scale is undertaken, there will be both scientific and political wrangling that will make our ears ache. Much of the debate centers on technical issues. The purpose of this chapter is to help you understand the technologies involved.

A first major step in assessing the global environmental issues will come in 1992, when 17 national space agencies will participate in the International Space Year, with the emphasis on what is called "Mission to Planet Earth." This project will be the first attempt to coordinate the mass of satellite-gathered information and to document all available details of global environmental change.

None of the early warnings of these environmental dangers is new. The *greenhouse effect* was first described by Dr. Roger Revelle, considered by many to be the dean of global ecology, back in 1957. Revelle expressed concern about the vast amounts of carbon dioxide released into the atmosphere by the burning of fossil fuels. Revelle wrote, "Mankind is inadvertently conducting a great geochemical experiment."

The threat posed by chlorofluorocarbons to the ozone layer, which shields the Earth from harmful solar radiation, was first proposed as a theory by F. Sherwood Rowland and Mario J. Molina, both at that time at the University of California at Irvine, in the summer of 1974.

Dire warnings about "death from the sky" were headline news in the 1970s, and acid rain became a much-debated issue. Some of the early predictions of environmental catastrophe from acid rain have proven to be overblown, but most scientists agree that this problem remains an important one.

If the warnings are not new, what is? Two developments have occurred in the past few years that place these early warnings in a new perspective. One, empirical data has been collected that supports the early warnings. Some indication of a global warming trend and the Antarctic ozone hole are two examples. Two, the assumption that Earth's climate would absorb shocks and respond to human influence in a steady, gradual way has come under question. Some experts now say that the climate can suddenly—within a century or less—flip into an entirely different mode. In fact, they argue that evidence recently drawn from polar ice and ocean sediments shows that past climates have already done just that.

Now take a closer look at the major environmental dangers and see what the evidence of their effects now indicates.

GREENHOUSE EFFECT AND CLIMATE CHANGES

Was the drought of 1988 attributable to the long-expected global warming trend known as the greenhouse effect? Was it a foretaste of climatic change that will by the next century cause unprecedented disruption in world environment? Or was it just a tough summer well within the normal range of climatic variability?

What do we know about the greenhouse effect, and why is one group of scientists telling us that the greenhouse effect is here now while another group challenges this theory and disputes the dire warnings? Take a look at the theory first and then see what the controversy is about.

Theory

If the Earth only absorbed radiation from the Sun without giving back an equal amount of heat, our planet would continue to grow warmer each year until the oceans and lakes boiled. Anyone who has been swimming in an ocean or lake lately knows that the water is not boiling, nor has the Earth's surface temperature exceeded historical variations. This state of affairs is known as energy balance, and it is crucial to our climate.

Climate on our planet is driven by the radiant energy received from the Sun. Radiation consists of electromagnetic waves traveling at the speed of light, 186,000 miles per second as you will recall from

Chapter 1. Light from the Sun, covering an average distance of 93 million miles, reaches us about 8 minutes later on Earth. We have to know a little about this radiation and what happens to it when it reaches Earth to be able to judge the greenhouse controversy.

Radiation is described in terms of its wavelength and frequency, that is, the distance between successive wave crests and the number of crests that arrive per second. When the wavelength is short, the frequency is high, and vice versa. For all forms of electromagnetic radiation, if you multiply any wavelength by its frequency, the result is the speed of light. The various forms of radiation compose the electromagnetic spectrum, as shown in Figure 4-1.

These forms of radiation range in wavelength from less than a billionth of a micron for gamma rays (a *micron* is a millionth of a meter) up to tens and hundreds of miles for long radio waves. From shortest to longest, the spectrum of waves includes: gamma rays, X rays, ultraviolet, visible light, infrared (heat), microwaves, VHF, television, and ordinary radio. The short ultraviolet rays are what tan and burn us.

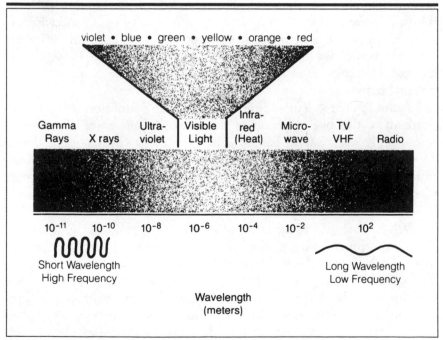

Figure 4–1 *Electromagnetic Spectrum*

Ordinary window glass does not allow these short rays to go through very easily, which is why humans don't get a tan through glass.

Sunlight is composed of all the colors of the rainbow. Each color—violet, blue, green, yellow, orange, red—has a discrete wavelength measured in microns. The shortest is violet, 0.4 microns; the longest is the color red, 0.7 microns.

Energy is received from the Sun in the form of light. The Earth absorbs some of this energy, which warms the planet's surface, but reradiates most of the heat energy back into the atmosphere as infrared (invisible but detectable as heat) at a characteristic wavelength of about 10 microns.

Climatologists explain that the continued burning of fossil fuel and other industrial and agricultural activities increase levels of carbon dioxide, carbon monoxide, methane, nitrous oxide, and chlorofluorocarbons in the atmosphere from 0.2 percent to 5.0 percent per year.

These gases absorb the infrared energy radiating from Earth's surface, thereby reducing the amount of energy that otherwise would be dispersed into space. The accumulating warmth is the essence of the greenhouse effect. Figure 4–2 illustrates this heat-trapping process. Thus, solar energy trapped in Earth's atmosphere could increase the average temperature by a significant amount over time. These higher temperatures could cause drought in some places, increase melting of the polar ice caps, raise levels of oceans, and disrupt worldwide agricultural patterns.

Although the greenhouse effect is attracting attention and concern these days, the phenomenon is not new. A variety of gases, most importantly water vapor, have warmed the Earth's surface for billions of years. Without these infrared-absorbing molecules in the atmosphere, the planet would be about 55 degrees Fahrenheit (30 degrees Centigrade) colder than it is today. It can be said that the greenhouse effect is, in fact, what made life possible on Earth.

What concerns some scientists today is a *runaway* greenhouse effect—a serious change in the balance of forces that control our climate. There is little debate about the theory, although the greenhouse analogy is not a perfect one. In a real greenhouse, the glass allows the solar radiation in and retains increased temperature primarily by preventing air currents from taking the heat away. The term as applied to the Earth's atmosphere refers only to the trapping of infrared radiation near the Earth's surface.

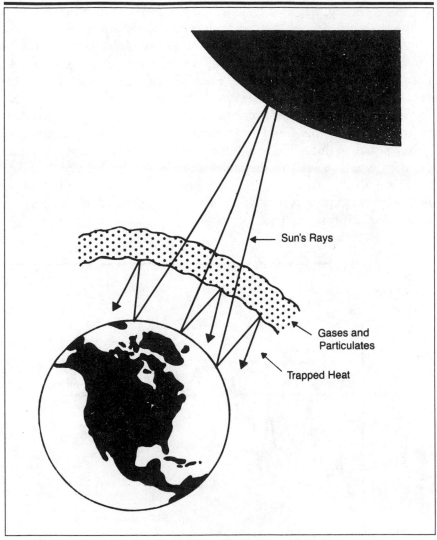

Sun's Rays

Gases and
Particulates

Trapped Heat

Figure 4–2 *Greenhouse Effect—Gases and Particulates Trap Heat Within the Earth's Atmosphere, Gradually Increasing Global Temperature*

Is It Getting Hotter, or Isn't It?

The uncertainty about the effects of fossil fuel emissions and industrial and chemical wastes raises the critical question of whether our planet truly is becoming warmer. This decade has seen the four hottest years of the last century, and the first six months of 1988 were the warmest on record. Figure 4–3 shows how the average global temperature in the 1980s compares with the average temperature for the previous 11 decades. According to this data, mean global temperatures have increased by 1 degree Fahrenheit since 1980 when compared to the 1950–1980 average (59 degrees Fahrenheit or 15 Centigrade). Again according to these figures, this trend is accelerating. The outlook is ominous enough that in June 1988 in his testimony before Congress, James Hansen, who is a leading climatologist at NASA's Goddard Institute for Space Studies, declared himself "99 percent" certain that the greenhouse effect is upon us now. If he is correct, we may look back to the summer of 1988 as the good old days.

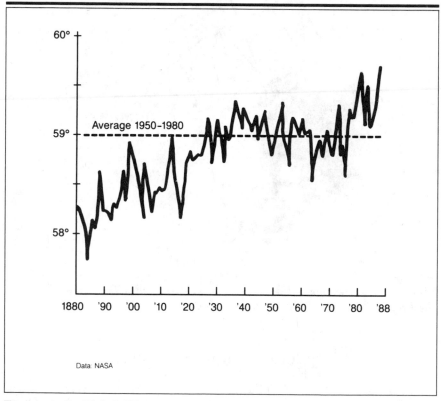

Figure 4–3 *Global Warming Trend*

Hansen is not alone. A large body of scientists attribute the apparent warming to the greenhouse effect. They point out that a century ago, carbon dioxide made up only 200 parts per million in the atmosphere. Today it is 345 parts per million and researchers expect it to double in the next century.

Carbon dioxide (CO_2) is only one of the gases in the atmosphere that help reflect infrared radiation back to Earth and thus amplify the greenhouse effect. The contribution of the other major constituents—chlorofluorocarbons, methane, and nitrous oxide—is shown in Figure 4-4. Chlorofluorocarbons are the gases emitted from plastic foams, fluids in air conditioners, refrigerators, and industrial solvents. Methane is released from a number of natural sources including cattle, bacteria in rice paddies, wetlands, and even termite mounds. Nitrous oxides come from both the burning of fossil fuels and the breakdown of nitrogen fertilizers. As can be seen from Figure 4-4, however, CO_2 is the major contributor to the greenhouse effect.

Where Does the CO_2 Come From?

As shown in Figure 4-5, the United States is the major producer of this gas. Within the United States, according to the Electrical Power Research Institute, power plants powered by fossil fuels account for 28 percent of all the carbon dioxide produced artificially. The other major producers are cars, trains, trucks, and planes, which emit 27 percent; and industry, which produces 29 percent.

Along with industrialization, the destruction of forests, especially in the tropics, also contributes to excess CO_2 in the atmosphere. Cutting down trees contributes to the greenhouse effect in two ways. First, the removed trees no longer absorb CO_2. Second, many of the fallen trees end up being burned and thus release CO_2 into the atmosphere.

Experts predict increased carbon dioxide and other gases that absorb infrared reradiation from Earth will increase the average worldwide temperature by up to two degrees Fahrenheit in twenty years and by another two degrees by the end of the next century.

If they are interpreting their data correctly, this means that the planet will be the warmest it has been in 100 million years, perhaps the warmest since life moved from the oceans to the land 400 million years ago. Great regional variations will occur. At the poles, where

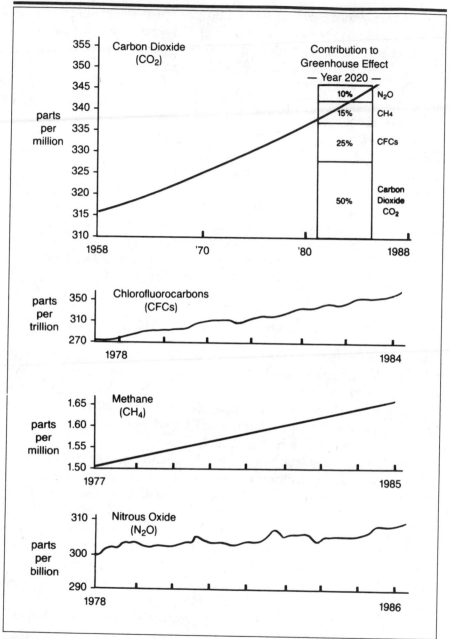

Figure 4–4 *Major Gases Contributing to the Greenhouse Effect*

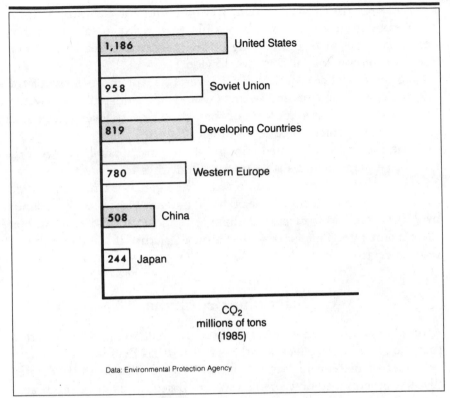

Figure 4-5 *Major Contributors to Global Warming*

temperatures have strong effects on worldwide weather, the rise in temperature may be as great as 10 degrees.

What's So Upsetting About a Warmer World?

We usually associate warmth with good, so superficially the idea of a warmer climate is appealing. In fact, not everyone would be a loser in the predicted climate change. Parts of the Soviet Union, Canada, and Scandinavia, for example, would enjoy a longer growing season. Net global damage, however, would far exceed local benefits. The corn and wheat belts in the United States would be parched as rainfall declines a predicted 40 percent. Roger Revelle, the past president of the American Association for the Advancement of Science, has calculated that a four-degree rise in average temperature could cut California's water supplies by more than half. This would turn half the state back into desert. As a result of the melting of arctic ice packs, sea levels could rise, threatening coastal cities worldwide.

As the global sea levels rise by several feet or more, most of the nation's coastal marshes and swamps would be inundated by salt water. Louisiana, which has 40 percent of the nation's coastal wetlands, could lose up to 85 percent of those ecologically rich habitats.

The EPA was concerned enough about this possibility to recommend that state agencies start now to find ways to protect America's coastal wetlands from the rising sea. The agency warned in a 1988 study that the nation could lose 30 to 80 percent of its coastal wetlands by 2100 if emissions of carbon dioxides and other gases continue at the current rate. This is altogether a bleak picture if the data is being read correctly—but is it?

Scientific Brouhaha

A number of qualified weather experts are challenging the prediction that the greenhouse effect is with us today. Reid Bryson, a prominent professor of meteorology at the University of Wisconsin, has stated that the greenhouse-now predictions are based on flawed data. If anything, Bryson and others claim, the worldwide climate over the past 40 years is getting colder, not warmer.

Even more outspoken is Kenneth E. F. Watt, professor of zoology and environmental studies at the University of California at Davis. He has called the greenhouse theory "the laugh of the century," and "a puff piece blown up in the media." Another critic is James Goodrich, one-time chief climatologist for the state of California. These and other critics point out that much of the data used by the greenhouse theorists is relatively new, and the samples are not large enough to predict long-term trends.

According to James Goodrich, the data being fed into computers and used to reach the alarming conclusions are skewed by questionable numbers collected under sometimes-questionable circumstances in unquestionably overheated cities. It is unscientific, Goodrich points out, to use urban temperature records for monitoring long-term climatic trends.

Kenneth Watt agrees and has pointed out that the *urban heat-island effect* has been well known since 1952. According to this theory, thermometer readings in or near cities are as much as 18 degrees Fahrenheit (10 degrees Centigrade) warmer than those in the surrounding countryside, with the degree of elevation relative to the population size of the city. Los Angeles, for example, has become 4 degrees warmer than the surrounding countryside in the last 60 years. According to Watt, all but 270 of the 11,600 weather stations in the National Weather Service in the United States are in cities or at airports near cities.

In a letter to *The New York Times* (August 11, 1988), Watt pointed out that the National Weather Service has identified a set of rural stations that are judged suitable for drawing inferences about climate trends. Almost all, he said, show markedly lower temperatures since 1940.

What are we to believe in the light of this scientific brouhaha? Senator Edmund Muskie made a plaintive call on the occasion of a Senate hearing on the health effects of pollutants. Testimony from scientific sources was not as definitive as the Senator desired. Witnesses insisted on saying, "on the one hand the evidence is so, but on the other hand. . . ." The exasperated Senator Muskie said, "What this country needs is a one-armed scientist."

In lieu of more definitive advice from the scientific community, our policymakers may be forced into a wait-and-see approach, but this too has its dangers. Climatologist Stephen Schneider of the National Center for Atmospheric Research in Boulder, Colorado, has used the analogy of an exceedingly dusty and murky crystal ball. Researchers are having difficulty in predicting the climatological future using this crystal ball, Schneider has said, but if they devote too much time to polishing this ball, the picture they eventually see in it may not be to their liking.

What Can Be Done?

Despite the maverick critics, much of the scientific establishment in the United States tends to accept the greenhouse theory and many, like James Hansen of NASA, are recommending immediate congressional action to slow the degradation of the Earth's atmosphere and set targets for global reduction in the burning of fossil fuels.

In a meeting in Toronto in mid-1988, government officials, scientists, and environmentalists from 48 nations called for a 20 percent worldwide reduction in the use of oil, coal, and other fossil fuels by the year 2005 and an eventual 50 percent reduction in the use of such fuels.

In the United States, 16 senators have unveiled a comprehensive plan to combat the global warming threat. "The greenhouse effect is the most significant economic, political, environmental, and human problem facing the twenty-first century," said Senator Tim Wirth, Democrat of Colorado, the principal author of the plan. The proposed legislation would force this country to reduce carbon dioxide emissions by 20 percent by 2000.

Wirth's proposal would require administration officials to draft a national energy plan. The plan would emphasize the least costly ways of reducing the use of fossil fuels and increasing the use of renewable energy such as solar power. A key element in Wirth's package is the requirement for new car models to average 55 miles per gallon by 2010, or double the 1985 fuel efficiency standard.

The proposed legislation calls for the Secretary of State to hold an international meeting aimed at persuading other nations to commit to a 50 percent cut in the gases that produce the greenhouse effect by 2015.

Senate Energy Committee chairman Bennett Johnston has said that in 1989, "We'll have to begin the serious process of legislating. We'll act as fast as the public's willing to support us." Thus, the politicians are waiting for you and me to voice sufficient concern on this issue, and the more we know about the issues involved the better advice we can give them.

THE OZONE WAR

On a personal level, I was a foot soldier in the ozone war since its opening shots. My role was modest, but I did have the opportunity to observe the major players in action—both the heroes and those whose actions were not so heroic. The scientists who sounded the first warnings and then came under heavy attack from affected industries and Washington bureaucrats deserve recognition. They were, among others, Harold Johnston of the University of California at Berkeley, the previously mentioned F. Sherwood Rowland, and Mario Molina. Serious efforts were made to distort their findings or hush them up, and I personally witnessed some of these.

The first warning that human activities could alter the amount of stratospheric ozone came in 1971 from Professor Johnston in connection with the supersonic transport aircraft (SST). A four-year study then ensued, sponsored by the Department of Transportation and involving scientists from all over the world. It was one of my tasks to write the final Report of Findings for that monumental study. As I witnessed politics starting to play an ever-stronger role influencing what should have been an impartial scientific effort, I became concerned. I wrote a paragraph for a section called Principal Scientific Conclusions that stated the concerns raised in 1971 about a large fleet of SSTs—concerns that were the impetus of the study in the first place—were fully justified. This paragraph became my personal litmus test as to the degree of political influence on the report. Although this statement appeared in several early drafts of the Report of Findings, it was dropped before final publication. I had my answer.

As in the SST case, the fluorocarbon/ozone issue quickly became politicized. As the principal investigator of a 1976 National Science Foundation study, I saw the chlorofluorocarbon (called CFC) chemical manufacturers and users mount major efforts to denigrate and minimize the warnings sounded by Rowland and Molina. These efforts were largely successful in fending off any meaningful government action until the recent discovery of the Antarctic ozone hole.

One of the main reasons for my views on the importance of a technologically literate society in this country stems from my experience in Washington, D.C. We cannot, in my view, leave environmental questions to the bureaucrats or the politicians.

For the past 50 years, industrial societies have been injecting potent chemicals into the atmosphere, unintentionally setting in motion the largest, longest, and most dangerous chemistry experiment in history. Now there is convincing evidence some of the chemicals (specifically chlorofluorocarbons) are destroying the ozone layer that encircles the Earth and shields it from ultraviolet (UV) radiation.

In our busy daily lives, we don't spend a lot of time worrying about the difference between the troposphere and the stratosphere. We may even be a little vague about where they are. A quick review of the basics will help put the ozone depletion issue into perspective.

The Basics of Earth's Atmosphere

Differences in the way temperature changes with altitude have led scientists to divide the atmosphere into distinct regions. Temperature falls with increasing altitude above the Earth's surface to a minimum value of about −80 degrees Fahrenheit (average between winter and summer) and then rises. The point at which the temperature reverses itself is called the *tropopause*, and its height varies from about 25,000 feet near the poles to about 50,000 feet in the tropical latitudes, as shown in Figure 4-6. The tropopause marks the boundary between the two lowest layers of the atmosphere. The main reason for the temperature inversion at this point is the ozone layer. The region

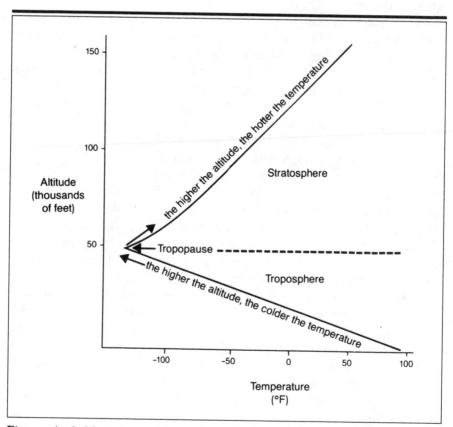

Figure 4–6 *Variations of Atmospheric Temperature with Altitude (Tropical Latitude 0 degrees)*

between ground level and the boundary of the temperature reversal is known as the *troposphere*—from the Greek meaning sphere of change. It is the region in which we live, and it is the area that contains winds, rain, storms, and other weather events. It is also a region of relatively rapid circulation and vertical mixing of air.

The stratosphere lies above the troposphere and here the temperature rises with altitude to a maximum value of 50 degrees Fahrenheit (10 degrees Centigrade) at about 160,000 feet. The stratosphere is a stable, virtually cloudless region with very slow vertical mixing of air.

The climatic differences between the two regions are important in terms of atmospheric pollution. In the troposphere, cold air is above warm air. The cold air is denser, so it tends to sink, while the warm air tends to rise. This vertical mixing of air is what causes various pollutants to be carried down to the ground and removed. Moreover, almost all precipitation occurs in the troposphere, and pollutants are thus rained out. The result of all this weather action is that the troposphere can usually cleanse itself of most pollutants in about a week or so.

In the stratosphere, on the other hand, the cooler air is at the lower altitudes and does not readily rise. There is no tendency for the air to mix. This stable situation is known as an *inversion* and is similar to the situation that occurs over Los Angeles and other natural basins, giving rise to serious urban pollution problems. Also, the stratosphere has no rain storms to wash out pollutants. Pollutants that would wash out of the troposphere in a week or so remain in the stratosphere for many years. A National Academy of Sciences study said, "The stratosphere can be likened to a city whose garbage is collected every few years instead of daily."

Ozone and the Ozone Layer

Ozone is a form of oxygen with three atoms (as opposed to normal oxygen molecules, which contain two atoms) that primarily exists between altitudes of 6 to 30 miles (10 to 50 kilometers) above the Earth. Ozone is created when ordinary oxygen in the stratosphere is bombarded by ultraviolet radiation from the Sun. Because it efficiently absorbs ultraviolet radiation, ozone protects life both in the lower atmosphere and at the surface of the Earth from the harmful effects of solar radiation. Because ozone plays an important and vital role in the atmospheric circulation and climate of the planet and in the viability of life, decreases in the amount of ozone could change climate and cause adverse effects that include increases in skin cancer and a weakening of the human immune system. If the fragile ozone shield is seriously damaged, the food chains of both land and ocean could be disrupted and all life on Earth would be at risk.

The Threat

Scientists see the main threat to the ozone layer as coming from the continued addition to the atmosphere of synthetic chemicals called chlorofluorocarbons (CFCs). Because CFCs are immune to destruction in the troposphere, their continued manufacture and release lead to ever increasing amounts in the stratosphere as they eventually float upward. In the stratosphere, CFCs are broken down by sunlight and the result is chlorine. Chlorine has a catalytic and destructive effect on ozone.

How can a relatively small amount of a man-made chemical affect something as large as the atmosphere? The answer is in that word, *catalytic*. A *catalytic chain reaction* is a series of two or more chemical reactions in which a substance (the catalyst) destroys another substance—in this case, ozone. In a process that is repeated over and over, each molecule of the catalyst destroys thousands of ozone molecules. In fact, a single chlorine atom can, over time, destroy 10,000 ozone molecules. The bottom line is that comparatively small amounts of pollutants can have devastating global effects.

What are these chemicals used for and where are they used? Primarily, two forms of chlorofluorocarbons are used: CFC-11 and CFC-12. In the United States, CFC-11 is mainly used in the production of both rigid and flexible foam products such as bedding, furniture, appliances, packaging, flotation, and fast-food containers. CFC-12 is primarily used in refrigeration and air conditioning. Industrial use by region (1985 data) is

United States—660 million pounds

Soviet Union and Eastern Europe—320 million pounds

Western Europe, Japan, Canada, Australia, New Zealand—933 million pounds

Developing countries—367 million pounds

As mentioned earlier, warnings of this danger were sounded in 1974 by chemists Sherwood Rowland, Mario Molina, and others. They were largely ignored and for 14 years their predictions were ridiculed and belittled by the chemical manufacturers, industrial users of CFCs, and their political supporters. Big money was involved. All told, U.S. industry sells $750 million worth of CFCs annually. Continually varied projections by the National Academy of Sciences relative to the amount of ozone loss added to the confusion. The result was a wait-and-see policy and the excuse not to take meaningful action to curb the production and use of CFCs.

The United States, Canada, Norway, and Sweden did ban the use of CFCs as a propellant in aerosol spray cans in 1975. Remember the great battle of the hairspray and deodorant cans of that period? In retrospect, this action had little overall effect, because the problem is worldwide—the atmosphere does not recognize national borders—and other uses for the chemicals increased, more than making up the difference.

The Hole in the Earth's Ozone Shield

Then, in 1985, came the discovery by a British Antarctic survey team of a winter hole in the ozone shield. (The hole becomes smaller or disappears in the Antarctic summer, when the temperature and winds change.) Prior to this discovery, ozone depletion theories were based on computer modeling, and empirical data was both in short supply and possibly explainable as natural variations. The ozone hole, however, was something entirely different. It was definitely there, it could not be explained by any natural variation, and it seemed to be worsening. The loss was measured at 50 percent in 1985 and 60 percent in 1987, when the hole was judged to be as big as the continental United States and spreading toward South America and Australia. Scientists have now found evidence that the Arctic suffers a similar but far less pronounced ozone loss during late winter.

Despite this evidence, the ozone war went on. CFC supporters claimed that the ozone loss was confined to the Antarctic. This last-ditch defensive position was abandoned by the CFC loyalists in March of 1988 when a new NASA study reported that atmospheric ozone over the Northern Hemisphere had declined significantly over the past two decades. The 1988 NASA study, considered the most authoritative so far, also found the springtime loss of ozone in the Southern Hemisphere was spreading into wider areas and that ozone levels were reduced throughout the year. (Figure 4–7 shows the decline in the ozone shield of 1.7 to 3 percent.)

The surprisingly rapid depletion of ozone now confirmed by satellites has prompted widespread rethinking of forecasts that the change would be gradual. And now comes the bad news. The conservative wait-and-see policy of the last 14 years carried with it a high price tag. Even if CFC emissions are halved in the next decade, as called for by an international treaty, levels of chlorine gas in the stratosphere will continue to rise.

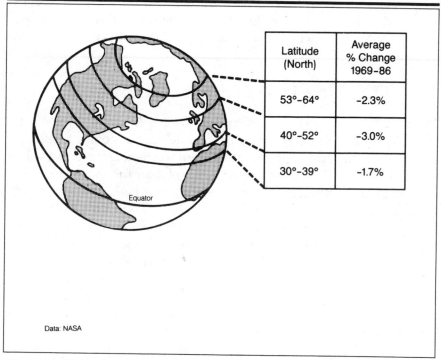

Latitude (North)	Average % Change 1969–86
53°–64°	-2.3%
40°–52°	-3.0%
30°–39°	-1.7%

Data: NASA

Figure 4–7 *Decline in Global Ozone Shield*

Figure 4–8 shows how long-lasting chemicals released in the mid-1980s will continue in the stratosphere for decades after the controls mandated by global treaty have been carried out. The harmful chemicals take 7 to 10 years to rise to the stratosphere, and once there, they persist for 75 to 100 years, destroying ozone all the while. Some 90 percent of the CFC molecules already in the air today will remain in the upper atmospheric regions in 2000 AD. We are going to have to live with our past mistakes; the ozone depletion will worsen. Our children and grandchildren will pay the price throughout the twenty-first and twenty-second centuries.

Take a look at the price of our failure to act when first warned about the danger. Ultraviolet radiation, particularly in the shorter wavelengths of the spectrum known as UV-B, is unquestionably the cause of most skin cancer. Each 1 percent drop in the ozone level allows 2 percent more UV-B to reach the surface of Earth, increasing the number of skin cancer cases by 3 to 6 percent. In the United States alone, the Environmental Protection Agency has calculated there will be more than 155 million *additional* cases of skin cancer and 3.2 million *additional* cancer deaths over the next century if the ozone destruction continues at its current pace.

Even worse could be the damage to other organisms. The effect of increased ultraviolet radiation on crops, mammals, fish, bacteria, trees, and grasslands has yet to be estimated. Humans can protect themselves from increased radiation by wearing clothes or by the use of protective sunscreen lotions, but plants and animals cannot.

What Can Be Done?

Phasing out all production of CFCs as soon as possible is crucial, but as you have seen, it may be too late even now. Even if all the producing nations agree to halt production immediately, the ozone layer will continue to diminish as a result of the CFCs already in the atmosphere. With environmental disaster looming, some scientists are dreaming up some pretty far-out technical fixes.

Among the fantastic countermeasures suggested is the use of giant lasers atop mountains to cleanse the atmosphere of harmful chemicals before they rise to the stratosphere. Dr. Thomas H. Stix, a Princeton University physicist, is the originator of this concept, which he calls "atmospheric processing." The laser would be tuned to the frequency

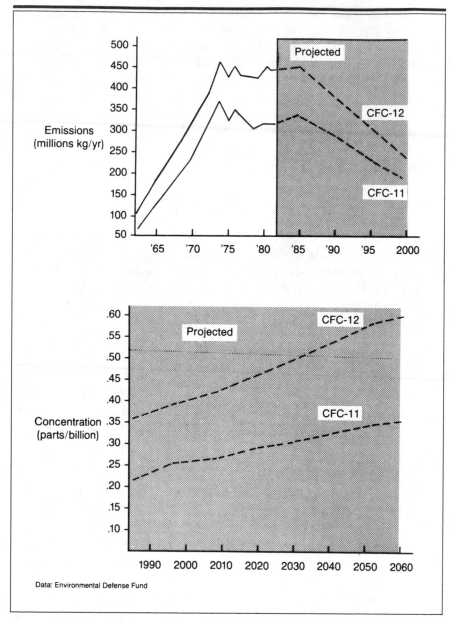

Figure 4–8 *Long-Lasting Chemical Threats to the Ozone Shield*

most damaging to the CFCs, which is in the infrared part of the spectrum. According to Dr. Stix, an array of infrared lasers around the world could destroy one million tons of CFCs a year. Futuristic? Yes, but is it any more futuristic than Star Wars missile defense systems?

Ozone replacement is another approach to the problem. Some experts have proposed that bulk ozone be produced on Earth and propelled into the stratosphere in rockets, aircraft, or balloons. Other ideas include placing solar-powered ozone generators in high altitude balloons or lofting lumps of frozen ozone into the stratosphere. Other scientists have calculated that replacing ozone is a massive task beyond our ability to accomplish today.

The main problem involves the energy needed to transport ozone to the stratosphere. The expert on the subject, F. Sherwood Rowland, explains that if we total all the energy humans use today, it does not equal the amount needed to reverse the ozone thinning.

There is a paradox between the stratospheric ozone layer needed to protect life on Earth ("good ozone") and the ground-level ozone that is the principal element of urban smog ("bad ozone"). When ozone levels increase down here in the troposphere, breathing can become difficult, especially for asthma victims, the elderly, and those who exercise. Actually, it's the same ozone, but what is good for us as a protective layer in the stratosphere is bad for us on the bottommost atmospheric level where the ozone molecule reacts adversely with our lungs and eyes.

Why not take ozone from our cities, where we don't want it, and somehow transport it up to the stratosphere, where we need it? In addition to the huge amounts of energy needed to do so, as Rowland has pointed out, there is the matter of relative amounts involved. On very bad days in Denver or Los Angeles, ozone levels can reach 0.3 parts per million for a few hours. Normal concentration of ozone in the midstratosphere is 30 times that amount. So even if we could somehow transport our urban smog up to the stratosphere, there just isn't enough ozone down here to make a difference.

ACID RAIN: THE FACTS AND THE FALLACIES

What is acid rain and why do experts announce one day that it represents the most serious environmental problem of the century, whereas a few months later experts announce that the problem has been overstated and may not be the widespread ecological disaster it was once believed to be?

The term *acid rain* covers a host of phenomena and is something of a misnomer anyway. Because of the occurrence of acid snow, acid sleet, acid hail, acid frost, and because dry deposits of acid particles can also contribute to the problem, scientists now prefer the term *acid deposition*. Here the more popular term, acid rain, covers the entire phenomenon.

Rain is naturally slightly acidic. Acid rain then is the term used to cover precipitation of higher-than-normal acidity. Acidity is measured on the pH scale, which runs from zero to 14, as shown in Figure 4–9. The lower the pH number, the more acidic the substance. The pH scale is logarithmic, so that water with a pH of 5 is ten times more acidic than water with a pH of 6, and water with a pH of 4 is one hundred times more acidic than water with a pH of 6. A measurement of 7 is neutral (neither basic nor acidic), with the numbers above 7 indicating increased alkalinity. Note that pure water is neutral, whereas natural rain is about pH 5.6. Vinegar is pH 3, lemon juice is pH 2, and battery acid is pH 1.

Rain is made unnaturally acidic by emission of sulfur dioxide and nitrogen oxides from the combustion of fossil fuels. (Yes, the same burning of the same fossil fuels that are the designated culprits in the greenhouse effect are to blame for this environmental threat.)

In the United States, rainfall is made more acidic by the 21 million metric tons of sulfur dioxide emitted from the burning of coal each

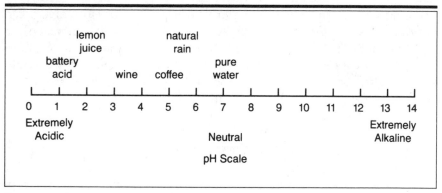

Figure 4–9 *Acidity and the pH Scale—The Lower the pH Number, the More Acidic the Substance*

year. Utility power plants in the eastern United States produce 70 percent of this pollution. Our rain is made more acidic by the 19 million metric tons of nitrogen oxides emitted from vehicles burning gasoline and oil, and to a lesser extent from other sources. Because of these and other pollutants, average annual rainfall east of the Mississippi is now about pH 4.5. Parts of Ohio and New York now average 4.1 pH and less. Acid clouds and acid fogs in the Los Angeles basin have been measured as low as 1.7 (more acidic than lemon juice).

Concern about the potential danger of acid rain led Congress in 1980 to fund a $500 million study of the issue called the National Acid Precipitation Assessment Program (NAPAP). The final NAPAP report is not due until 1990, but preliminary findings seem to downgrade the original fears of 1970. The following summarize these findings:

- No significant reduction in growth of food crops or tree seedlings has been found.

- Damage to forests—defoliation and dead or dying trees—has occurred, but whether this can be completely attributed to acid rain is not now clear.

- In certain sensitive areas (the Adirondacks, for example) many lakes and streams have become acidified because of acid rain, and their fish life has vanished. That has not been true of the lakes and rivers of the relatively lightly industrialized West. It is the prevailing jet streams that carry pollutants from power generating plants in the Midwest and West to the Northeastern United States and to Southeastern Canada.

- No adverse effects on humans have turned up.

Does this mean you can stop worrying about acid rain? Unfortunately, no. First, the NAPAP report is only interim and findings could change. Second, there is plenty of bad news in the interim report. There is no doubt, according to NAPAP, that where sulfur emissions from industrial smokestacks have increased, sulfates in surface water also have increased. How much damage this causes to lakes and streams depends on the natural acidity of the water and the species of fish.

Tests have shown that highly acidic rain—pH 3.0 or lower—does damage some crops. Natural rain is, as I pointed out, pH 5.6, and average rain acidity in the U.S. ranges between pH 4.1 and 5.2. Thus, as far as crops are concerned the level of acidity is acceptable for the present. The problem is that the acidity levels are worsening each year.

In summation, although short-term catastrophic effects of existing levels of acid rain have not appeared, the long-term effects remain in doubt. As a nation, we may have time to take the necessary steps to head off serious trouble.

Corrective Action

As you may have read, Canada is quite concerned about the acid rain problem, particularly about the area north and east of Toronto. Canadian scientists say that 75 percent of acid rain in Canada comes from the United States. Canadians are lobbying the U.S. government to enact legislation to cut emissions by 50 percent of the 1980 levels by the mid-1990s.

How are these cuts to be made? The most effective way of controlling pollutants that cause acid rain would be to burn less fuel for energy generation and transportation. More fuel-efficient cars and expanded mass transport systems can reduce oil consumption for transportation, but energy generation is a more difficult problem. In spite of worthy exhortations to conserve energy, demand is more likely to increase in the future.

Reduction of emissions from fossil fuel-burning power plants is the key to reducing acid rain. Coal-burning plants are the main sources of pollutants, and it may be necessary to repower (that is, refurbish with new combustion systems incorporating clean-coal technologies) all of our aging coal-burning power plants. Clean coal is coal naturally low in sulfur or coal that has been washed to remove sulfur and other contaminants. This approach could cut sulfur dioxide emissions by more than 80 percent and nitrogen oxide emissions by more than 50 percent. Unlike stratospheric ozone depletion, the acid rain problem is still correctable.

WASTE DISPOSAL—YOUR BACKYARD OR MINE?

The world is running out of places to dispose of its waste—ordinary refuse and household garbage, toxic industrial waste, and nuclear waste. The result is that the dangers of waste materials to human health are increasing rapidly. We are busy fouling our nest every day of the week, and the horror stories abound—from the loathsome fetid rain of Mexico City to the tainted beaches of New York and Los Angeles—we are facing a major worldwide waste disposal problem.

Household Waste

Our throw-away society, along with other developed and industrialized societies, generates the most trash globally. However, the developing countries have waste disposal problems as well, because they lack the advanced technologies to deal with the problem.

A citizen in New York City, for example, generates four pounds of garbage per day, compared to a citizen of Kano, Nigeria, who, it is estimated by the World Bank, generates about one pound of garbage per day.

We admire Japan's ability to produce goods, but it turns out that we may also have much to learn from the Japanese about the disposal of solid waste. The Japanese have shown what can be done, and their waste disposal figures put our own into some perspective. The Japanese put at most 20 percent of their unprocessed household waste into landfills, whereas we in the United States dump 90 percent of ours into landfills. Unlike most U.S. landfill sites, Japanese dumps use impermeable liners, leachate collectors,* and waste-water treatment to prevent pollutants from escaping into groundwater.

Recycling is another area in which Americans fall behind other nations in waste disposal. Again, Japan can be a model. For example, Japan recycles 50 percent of its paper, and 95 percent of its beer bottles are reused an average of 20 times. The United States recycles only about 25 percent of paper used, and only 7 percent of glass is recycled. After recycling, only half as much waste per person is generated in Japan as in the United States. Much of what is not recycled in Japan is incinerated in facilities that produce energy as a byproduct.

How can Japan, with a limited land area, high population density, and scarce resources, manage its waste disposal problems so well? One answer may be in the close coordination among local, regional, and national government agencies in the collection and disposal of waste.

---

* Leachate is the soluble constituents of a landfill that, if not collected, percolate down to contaminate the water table.

In this country, garbage collection and disposal are local issues, but we may not be allowed this autonomy much longer. The United States may in time be forced to consider waste disposal as a national problem worthy of national attention.

Toxic Waste

What are toxins and how can they hurt you? *Toxic* is another word for *poisonous*; almost anything is toxic if taken in a large enough dose. As the term is used in environmental issues, *toxins* means substances that are poisonous in very small amounts, often measured in parts per million. Parts per million are difficult to imagine unless you compare this measure to items with which you are more familiar. One part per million is equivalent to

one inch in sixteen miles

one ounce in 31 tons

one tablespoon in 4000 gallons

Toxic substances can be harmful whether they are eaten, inhaled, or absorbed through the skin. They can damage the body in many ways. A substance is called a *carcinogen* when it causes cancer, a *mutagen* if it causes genetic damage, or a *teratogen* if it causes a fetus to be malformed.

According to the Environmental Protection Agency, American industry is pouring more than 22 billion pounds of toxic chemicals into the air, water, and land each year. These figures were supplied to the EPA by chemical companies and other industries under a new "right to know law," and the statistics cover the year 1987. The number of pounds of toxic substances took government officials by surprise. "The numbers are startlingly . . . unacceptably high and far beyond what we thought was occurring," said Linda Fischer, assistant EPA administrator for policy and planning at a news conference in April 1989 when the figures were released. How the 22.5 billion pounds of toxic releases broke down is shown in Table 4-1.

The chemical manufacturing industry was the biggest source of toxic pollution, with 3849 plants nationwide emitting 12 billion pounds or about 56 percent of the total release. Paper mills and factories dis-

Table 4–1 Toxic Chemicals Released by Major American Industries in 1987

AMOUNT RELEASED (IN BILLIONS OF POUNDS)	DESTINATION
9.7	Streams and other bodies of water
2.7	Air
2.4	Landfills
3.2	Deep in the ground
1.9	Municipal waste water treatment plants
2.6	Offsite treatment and disposal facilities
22.5	Total billions of pounds of toxic pollutants

Source: EPA data released in 1989.

charged 2.8 billion pounds, and primary metal plants, such as smelters and steel mills, released 2.6 billion pounds.

Disposal of this toxic waste is proving to be a technically difficult problem. Burying toxic waste is not an effective manner of disposal. Buried waste has a tendency to leach into the underground water supply with serious contamination resulting. Why not just burn it, then? Incinerating toxic waste is much more expensive than burning ordinary waste. It must be done at high temperature, the gases produced must be collected, and the resulting ash must be disposed of. In short, burning toxic waste only generates new forms of toxic waste. If not burial or burning, then what works?

A whole new industry has started based on developing new technologies to render toxic waste harmless. The Environmental Protection Agency is encouraging development of innovative methodologies such as a *vacuum extraction process* that is designed to minimize the cost and risk of incineration at sites with toxic wastes that evaporate easily, such as petroleum products or solvents. The process works on the same principle as a vacuum cleaner. Holes are drilled in contaminated soil, and the vacuum equipment sucks out the toxic vapor. The vapors are then incinerated but the volume is now much smaller than the large amounts of contaminated soil that would otherwise have to be disposed of.

Biodegradation is another approach that could prove useful. It consists simply of adding microbes that eat the waste, and oxygen or water and nutrients to make the microbes work faster. But biodegradation has limitations. The biodegradation process can be slow and is generally limited to locations that have one predominant organic waste. This is because microbes eat organic compounds, and mixtures of chemicals could include those that are toxic to the microbes.

Several biotechnology companies are trying to develop tailor-made microbes to work on specific toxic wastes. General Electric has patented a genetically engineered microbe that breaks oil down into relatively harmless components. As I said in Chapter 2, genetic researchers hope to be able to use this approach on oil spills, but it has yet to be tested in the field.

New types of incineration are also being tested. One new process uses infrared heat to destroy the wastes, which are moved through the equipment using a mesh conveyor belt, providing more control over the operation and keeping the waste in the heat for a longer period to extract a greater proportion of the toxic chemicals.

These new approaches illustrate the public's and government's growing concern. The government has set up an $8.5 billion toxic dump clean-up program, and we can expect innovative techniques including thermal, chemical, and biological methods for destroying dangerous wastes.

Nuclear Waste Control

Nuclear waste is the problem that won't go away. Three decades after the first nuclear power plant went on line, and four decades after the explosion of an atomic bomb over Hiroshima, we are still searching for a place to store the byproducts of the nuclear age. There are 108 nuclear power plants in the United States; accumulating in large water pools at these plants is the dangerous stuff nobody knows quite what to do with.

Why is nuclear waste (sometimes called "radwaste") especially dangerous? Life on Earth has always been exposed to natural radioactivity. However, it is the concentration of radioactive substances in amounts far exceeding any found in nature that poses the threat to life. Radioactivity is the tendency of an atomic nucleus to decay through the emission of particles. Three emissions occur: alpha particles (each one being two protons and two neutrons), beta particles (each one being an electron or a positron), and gamma radiation. In sufficient strength, these emissions are harmful—even lethal—to living animals.

To date, the United States has accumulated 15,000 metric tons of so-called spent fuel at power plant sites. The Department of Energy estimates that by the year 2000, that figure will reach 50,000 tons. So what exactly is "spent fuel"? Uranium assemblies are removed from commercial reactors after three years' burnup. They are not removed because the radiation is "spent," but rather because they have become *too radioactive* for further efficient use. The term *spent fuel* is a gross misnomer.

Since 1945, when the first nuclear bombs were being made, billions of dollars have been spent to find ways to get rid of nuclear waste. Some of the options have included shooting the waste into space, depositing the waste on some remote island, burying it on the ocean floor, or leaving it in the Antarctic. None of these solutions, for political or practical reasons, has proved viable.

The current plan is to bury nuclear waste from nuclear power plants at one site, southern Nevada's Yucca Mountain. The plan is to dig into Yucca Mountain and create a repository. Once filled to capacity with 70,000 tons of waste, which will occur around the year 2030, the repository will be sealed off and its entrance shafts refilled. The idea is that the sheer bulk of Yucca Mountain and the artificial barriers of the repository will isolate the waste and protect the surface environment. Let's hope so! The waste to be placed in this stone vault will remain radioactive for millions of years. Will the repository be strong enough? Are there means of preventing leaching into underground water sources there? What about earthquakes? The Nuclear Regulatory Commission assures Americans that all is well and not to worry. Physicists hedge on their reassurance. They say it is stretching science beyond its limits to project the radiation doses the mountain may emit 10,000 years in the future.

Military high-level radwaste is also accumulating from the manufacture of the two to three nuclear weapons American manufacturers turn out every day of the week. Military radwaste has been accumulating at the rate of about 200,000 cubic feet a year for the past decade. There is some evidence that there are dangerous levels of radioactive pollution at all 16 of the Department of Energy's weapons laboratories. It is one of the great ironies of our age that in the name of national security, we may be poisoning ourselves.

The Department of Energy plans to store nuclear waste byproducts of the nuclear weapons program underneath the New Mexico desert, but regulatory and safety problems have delayed the planned October 1989 opening.

Called the Waste Isolation Pilot Plant (WIPP), the DOE facility contains 56 rooms carved out of salt deposits located 2100 feet below ground near Carlsbad, New Mexico. The Department of Energy plans to start placing waste in the facility next year as part of a five-year-long series of tests designed to demonstrate compliance with safety standards.

CONCLUSION

I said at the onset that this was not going to be a gloom-and-doom chapter, so I must end by confessing that I am a technical optimist. I believe that the problems that technology has caused for our environment, if we confront them, will be solved by technology. Our clear duty is to keep ourselves informed.

There are no easy answers to any of the environmental problems raised in this chapter. The CFC/ozone issue is one example of where our existing social, political, and scientific institutions proved inadequate to cope with a technical environmental question replete with uncertainties.

Is it possible that we now need a cabinet-level environmental advisor to the U.S. president? The 1989 Congress started to pay serious attention to environmental problems with the introduction of several major omnibus greenhouse bills. In the Senate, three pieces of legislation have been introduced, by Senators Gore, Wirth, and Leahy. Senator Gore's bill effectively requires automobile manufacturers to raise the average fuel efficiency of their fleets to at least 45 miles per gallon by the year 2000. Senator Wirth's bill requires the development of a national energy plan designed to reduce carbon dioxide emissions by 20 percent by the year 2000. The centerpiece of Senator Leahy's bill is a provision that would make carbon dioxide a pollutant under the Clean Air Act and would mandate substantial, phased-in reductions of CO_2 emissions by fossil-fueled power plants and vehicles.

Should the United Nations shift its emphasis from trying to prevent wars to protection of the shared planet? With the cold war about over (Japan and West Germany won, in case you haven't been paying attention), it may be that the major challenge of the 1990s will be how to maintain a habitable planet. The technologically literate can play a meaningful role in this worthy endeavor.

═══ KEY CONCEPTS ═══

▶ Emissions of carbon dioxide and other gases produced by the burning of coal, wood, and oil let in the Sun's light but trap the resulting heat as it tries to return to space. The result is called the greenhouse effect, which causes global warming.

▶ Climatologists argue about whether the greenhouse warming is changing our climate now, or if the warming will not arrive for several decades. There is agreement, however, that unless the burning of fossil fuel is curtailed, the Earth's atmosphere will eventually start to heat up.

▶ There is now convincing evidence that the release of chlorofluorocarbons into the atmosphere is thinning the ozone layer that protects living creatures from harmful ultraviolet rays.

▶ Acid rain encompasses all types of precipitation whose acidity is higher than normal (pH 5.6). Rain is made unnaturally acidic by emissions of sulfur dioxide and nitrogen oxides from the burning of fossil fuels—primarily coal.

▶ Although short-term catastrophic effects of existing levels of acid rain have not appeared, the long-term effects remain in doubt. The United States may have time to take the necessary steps to head off serious trouble.

▶ Toxic industrial waste has proven difficult to dispose of, because neither burying nor burning solves the problem. Developing new technology to dispose of hazardous chemical wastes safely is the key to the government's clean-up program.

▶ Disposal of nuclear waste is the problem that won't go away. Despite 30 years of reassurance from the nuclear power industry and the nuclear weapons manufacturers, no agreed-upon method for the safe disposal of dangerous waste exists.

5

New Rules
for the
ENERGY GAME

EXPERTS TELL US that energy—its generation and use—may well be the central issue of the 1990s, so careful attention to our national energy needs is warranted. Is there trouble ahead, and if so what are our alternatives? The aim of this chapter is to answer these questions reasonably. When the energy policy debates get into high gear sometime soon, we are all going to be set upon by True Believers of this or that single remedy. Nuclear power, solar energy, conservation—all will have their evangelistic adherents, and we will be hard put to sort it all out.

To prepare for the onslaught of energy debates, it's best to clarify, in general terms anyway, where we get our energy and how we use it. Also, you should know whether there are any major changes on the horizon and what these changes could mean to you and your life style.

FUNDAMENTALS

Because energy is used in many forms, with different physical qualities and different capacities, it is difficult to specify quantities in a common unit. What layperson can keep all the barrels of oil, gigawatts of electricity, tons of coal, and cubic feet of natural gas straight? To help make comparisons, this chapter uses common units wherever possible.

One common unit is the Btu or British thermal unit. A *Btu* is defined as the amount of energy required to raise the temperature of 1 pound of water 1 degree Fahrenheit (5/9 degree Centigrade), or specifically, from 39.2 to 40.2 degrees. A barrel of crude oil, for example, contains about 5.8 million Btus. When very large amounts of energy are discussed, it is convenient to use a unit called a *quad*, defined as a quadrillion (10^{15}) Btus. I know that your monthly electric bill is not computed in quads, nor do you buy your automobile fuel by the barrel, so frequently in this chapter the big numbers are converted to something we can all relate to. A quad is the energy contained in 8 billion gallons of gasoline, a year's supply for 10 million autos. Table 5–1 shows the current big numbers for the United States in 1987.

We are prodigious users of energy in the United States, and we are still shockingly wasteful of our energy resources. On an energy/GNP comparison basis—a measure of energy use efficiency—the United States ranks fifth in the world. Japan's usage is twice as efficient as that of the United States. When energy was cheap and we were a net exporter of energy, our profligate ways were not that important.

How important is our comparative efficiency now? Consider the following. The United States currently spends about 11.2 percent of its gross national product on energy, whereas Japan spends 5 percent. This relative inefficiency in fueling its energy needs costs the United States $220 billion a year and gives the Japanese about a 6 percent economic edge on everything they sell—both in the United States and

Table 5–1 Where We Get Our Energy

SOURCES	STANDARD UNITS	QUADS	PERCENTAGE
Petroleum	5940 million barrels	32	43
Coal	829 million short tons	17.3	24
Natural gas	17 trillion cubic feet	16.5	22
Nuclear	414.6 billion kilowatt-hours	4.5	6
Hydro	338 billion kilowatt-hours	3.5	4.75
Geothermal and other	6.6 billion kilowatt-hours	0.2	0.25
TOTAL		74	100

Data Source: *U.S. Statistical Abstract—1988*

in foreign competition with U.S.-made products. As you will see in the next section, the United States has made significant progress in reducing energy use, but there is still room for improvement. General ways that we in the United States use our energy are broken down in Table 5-2.

What Do the Numbers Mean?

The breakdown of usage in Table 5-2 gives us a pretty good big-picture view of our national energy consumption. More important, though, is the gap between what we as a nation produce and what we consume, as shown in Figure 5-1. The difference between what we produce and what we use is made up by imports (primarily oil). As Figure 5-1 shows, imports reached a peak in the 1977-1979 period, dropped off as we became energy conscious, and are now starting to rise again. Of the 18.4 quads of oil the nation consumed in 1986, the United States imported 8.8 quads. That's not so good and it is part of what is known as the energy problem.

The emphasis in this book is on technology, not economics, but the experience of the oil crises of 1973-1974 and 1979 dramatically demonstrated the problems of our dependency on unstable foreign governments for our energy supply. When the price of oil increases, which it will inevitably do, the United States has to do more work—that is, produce more goods and services—to buy a given quantity of

Table 5-2 How We Use Energy—1987 Data

END USE	QUADS	PERCENTAGE OF TOTAL
Electricity generation	26.8	36.3
Transportation	20.7	28
Industrial and miscellaneous	16.8	22.6
Residential and commercial	9.7	13.1
TOTAL	74	100

Data Source: *U.S. Statistical Abstract—1988*

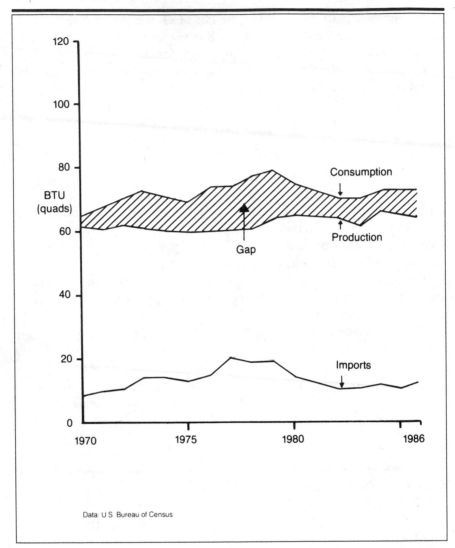

Figure 5-1 *Energy Production and Consumption—Gap Must Be Made Up by Imports*

oil. The large payments to foreign governments contract domestic demands for goods and services, and in effect become an excise tax—a
drag on our economy. In addition, U.S. oil imports are a significant
contributor to our continuing trade deficit. A growing trade deficit

results in higher unemployment, diminished opportunity, and a lower standard of living for most of us.

Experts and politicians began to recognize this problem about 15 years ago. The oil embargo of 1973, the natural gas shortage of 1976–1977, and the long lines at the gasoline pumps in 1979 dramatized the situation. As these symptoms of the problem were relieved, the general public thought the problem had gone away. It hasn't. Oil dependency is the name of our malaise. We are hooked on cheap oil. Much of our life style depends on cheap fuel, and this dependency is not going to be an easy habit to kick.

Oil prices vacillate in the short term and oil gluts alternate with oil shortages, but the longer-term future has never been in much doubt. According to researchers at New Hampshire's Complex Systems Research Center (whose findings were published in their book, *Beyond Oil*), the future is not a rosy one. Consider the following:

- U.S. oil supplies will be virtually exhausted by 2020. We won't literally "run out of oil," rather it will be just too expensive to obtain.

- As early as 1995, it could take more energy to explore for oil in the United States than the wells will produce.

- U.S. agriculture has grown so dependent on oil and its derivatives for every phase of the agricultural process from planting to harvesting to marketing that unless an economic substitute is found, the United States may lose its ability to export food in the next 20 to 30 years.

- World oil extraction is expected to begin to decline early in the next century. Canadian oil production will begin a steep decline in about 15 years, and U.K. production will decline in about 20 years. By that time OPEC may be a smaller, more cohesive group with much greater leverage than they have today.

Why We Must Consider Alternative Energy Sources

There are two main reasons to consider alternatives to oil and coal in the near future. First, our dependency on foreign energy sources seriously damages our economy and it is bound to get worse. Second, if the greenhouse effect is really pushing temperatures up to damaging levels, the world—and most particularly the United States—may have to limit the burning of fossil fuel.

Of all the fossil fuels, coal produces the most CO_2 per kilowatt hour of electricity, and given the scientific consensus on the seriousness of the greenhouse problem, the continued use of our abundant supply of coal may have to be curtailed.

Concerned about this, an electrical utility in Connecticut came up with an unusual response: The company will pay to have trees planted in Guatemala. This company, Applied Energy Services, contributed $2 million toward planting trees over 385 square miles in Guatemala. The number of trees planted will be able to absorb at least as much carbon dioxide as is emitted by the company's 180-megawatt generation plant in Connecticut, which will emit an estimated 15 million tons of carbon dioxide over its 40-year lifetime.

Is this type of balanced consumption and restoration the solution to the greenhouse effect? No, unfortunately, while an interesting and imaginative experiment, the Connecticut approach cannot be expanded to solve the entire coal-versus-greenhouse-effect problem. An estimated 6 billion tons of carbon dioxide is entering the atmosphere each year. At the same time, 26 million acres of trees, which would otherwise absorb some of the carbon dioxide, are being cut down annually. Tropical deforestation is taking place in Brazil, Indonesia, Colombia, and other places in the world at a rate considered detrimental to the ecological balance.

To compensate for that much carbon dioxide and that many trees being cut down, it would be necessary to plant as many as 3 billion acres of trees. That is an area bigger than the United States. Still, the Connecticut electrical company may be on the right track.

Whether the need to change to alternative sources of energy comes in the next 10 or 30 years, no one doubts that the era of cheap oil and plentiful coal will end. When that times comes, this country's economy and our life style will be endangered.

The Need for a Balanced Energy Budget

The solution to the problem is not simply to produce more energy by different means, and not simply to conserve energy use, but rather to find a new equilibrium between supply and demand. As Mr. Micawber in David Copperfield said, "Annual income twenty pounds, annual expenditure nineteen nineteen six, result happiness. Annual income twenty pounds, annual expenditure twenty pounds ought and six, result misery."

Apparently, if we are to balance our energy budget, both nuclear and solar energy sources will play a larger role in our future, as will many new ideas and inventions. Wind-driven machines, biogas digesters, and other renewable energy technologies may become practical and common sources of energy.

We are clearly on the edge of an energy transition era. The problem is in effecting a socially acceptable and smooth transition from gradually depleting resources of oil and natural gas to new technologies whose potentials are not now fully developed. The question is whether we are smart, diligent, and lucky enough to make the inevitable transition an orderly and smooth one. This chapter briefly reviews the major alternatives to our oil/coal dependency:

- improved energy efficiency
- nuclear power
- solar power
- exotic energy sources

First, I examine energy efficiency as a means of buying time for our eventual transition to new energy sources.

ENERGY EFFICIENCY

In general, throughout our economy it is now a better investment to save a Btu than to produce an additional one. You may wish to cut that statement out and fix it to your refrigerator (which, not so incidentally, is one of the largest users of electrical energy in your house). A word is in order here about the term *energy conservation*. To some people it has a bleak, puritanical connotation that implies a necessary decline in living standards—being too cold in the winter and too warm in the summer, or driving less than we want to. It doesn't have to be that way. We have learned about the efficient use of energy since the oil shocks of the 1970s, and one of the things that we have learned is that energy efficiency pays off.

Americans have enjoyed a 35 percent rise in gross national product since the early 1970s without increasing their energy consumption and without sitting in the dark and shivering. What is true in the United States has also been true for the rest of the world. In fact, the world has saved far more energy since 1973 through improved energy efficiency than it has gained from all new sources.

The industrialized world simply became energy conscious in the early 1970s. Better insulated buildings, more fuel efficient transportation, more efficient power generation and transmission all contributed to the energy savings.

If we were using energy in the United States at the usage rate of 1973, we would need 35 percent more fuel than we actually use. This difference in terms of oil and gas is equivalent to 13 million barrels of oil per day, or about half the entire production capacity of OPEC. To put this enormous savings in terms of dollars, conservation is saving this country $150 billion per year in energy costs, a total that is close to the size of the annual federal budget deficit.

So we have been doing well in the energy conservation department. Between 1973 and 1985, most Western European nations reduced their energy consumption by 18 to 20 percent. In the United States we reduced energy consumption by 23 percent, whereas Japan achieved a 31 percent reduction. This is all good news, but many experts say there is still room for improvement.

How Much More Can We Achieve by Energy Efficiency?

According to many experts, inefficiencies may still account for 50 percent of the energy used in the United States and most European countries, somewhat less than 50 percent in Japan, and more than 50 percent in the Soviet Union.

Possibly a better perspective can be achieved if these percentages are converted to dollars. In 1985, this country spent $440 billion on energy. This amounted to $5000 per household or, as I said earlier, about 11 percent of our GNP. If stringent conservation measures were taken and the United States became as efficient as Japan, for instance, we would be consuming half as much energy as we do today. We would save $220 billion per year (in 1985 dollars) and the cost of achieving this goal would only be about $50 billion. A good investment, indeed. But how do we go about achieving this amount of energy reduction?

Superinsulation

Superinsulated homes and commercial buildings offer the largest potential for energy savings. As many of you know from personal experience, insulating your home pays off in reduced heating and air-conditioning bills. Insulation is measured in R *values*, which is a measure of resistance to heat flow. The higher the number, the better the resistance. For instance, a standard attic ceiling is R-19, and a standard four-inch-thick insulated wall is R-11.

A superinsulated house would be built with heavily insulated walls and ceilings that offered ratings of up to R-30 for the walls and R-60 for the ceilings. The result would be remarkably low heating bills. A few hundred dollars in annual savings would pay for a lot of insulation over the lifetime of an average house. You have to do your own home's calculations individually—taking the cost of added insulation versus the amount you can save depending on your utility rates—but the calculation is worth doing. In a typical northern U.S. climate, if the thermal resistance of an average house is doubled, annual energy consumption is cut by about two-thirds.

Superwindows

About a third of the heat in the average U.S. home today escapes through windows. Windows are great for letting light in or for letting you look out, but they do not insulate well at all. The standard insulated wall has an R value of 11, but the typical single pane of glass window has an R value of only 1. Double-glazed windows have an R-2 rating, and they can be upgraded to R-4 by coating the inner surface of one pane with infrared radiation reflector material, such as tin oxide, and filling in the space between the panes with argon gas. Such windows are now commercially available in the United States, as shown in Figure 5-2. Argon, which is inserted at the factory, is a colorless, odorless, harmless, and chemically inactive gas currently used in fluorescent lamps.

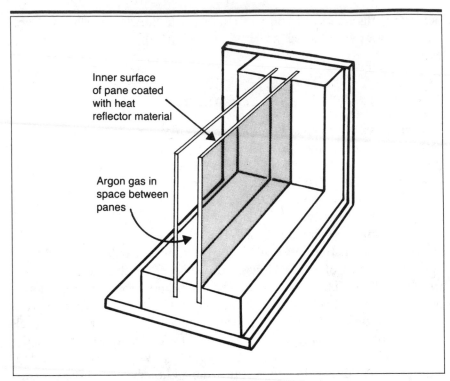

Inner surface
of pane coated
with heat
reflector material

Argon gas in
space between
panes

Figure 5-2 *Superinsulated Window*

Even better are triple-glazed windows, which have an insulating value of R-10 or R-11. In these superwindows, each inner surface of the two outer panes is coated with infrared reflectors and the space between panes is filled with krypton—a gas 60 percent less conducting than air and twice as insulating as argon. These superwindows will also become available commercially in the near future. When these windows are in common use in the U.S.—possibly sometime after the year 2000—they will save energy equivalent to 317,000 barrels of oil per day.

Commercial buildings offer a like opportunity for energy savings. Efficient building design can have a dramatic impact on heating costs. The worst of the older buildings in this country—those built before 1960—consume 270,000 Btus of energy per square foot. Such inefficiency means that over the 50-year lifespan of a building, the energy costs could double or triple the construction costs.

Because of rising fuel costs and because it is cost-effective to do so, future commercial buildings will be built with a careful eye to energy conservation. By using all available energy-saving technology, consumption of only 100,000 Btus per square foot could become standard.

Automobile Fuel Efficiency

Personal automobiles account for half of the energy used for transportation and about one-third of total petroleum consumption in this country. According to energy analysts at the Lawrence Berkeley Laboratory, an improvement of 0.1 mile per gallon in the U.S. automobile fleet would save the equivalent of 20,000 barrels of oil per day—roughly the estimated production rate from drilling proposed in the environmentally sensitive Georges Banks fishing grounds off the eastern United States.

I discuss the specific technologies involved in improving fuel efficiency in our private automobiles in Chapter 8, but for now, suffice it to say that significantly greater economies are technically possible, and a 60-mile-per-gallon (MPG) car is achievable today with no great technical breakthroughs.

More Efficient Lighting

We are currently using about 20 to 25 percent of U.S. electricity to provide illumination. New efficient fluorescent light fixtures use roughly 75 percent less energy than typical incandescent bulbs with the same light output. Figure 5–3 shows how much difference just one light fixture change would mean. There are an estimated 2.5 billion light bulb sockets in use nationwide. Of these, an estimated 1 billion could use the new screw-in high-efficiency fluorescents. Experts estimate that the widespread use of these high-efficiency fluorescents could save 500 billion kilowatt hours (kwh) of electricity between 1990 and 2010. Because about half of this would have been generated from oil and gas, the savings would amount to the equivalent of 560 million barrels of oil.

More Efficient Refrigerators

Refrigerators consume about 7 percent of the nation's electricity. New refrigerators, which started coming into the marketplace in 1985, use only about half of the kilowatt hours that the old ones use. Refrig-

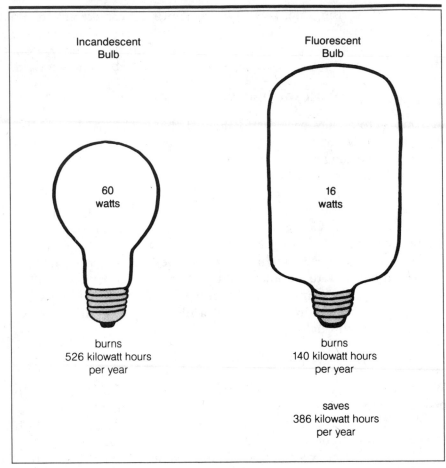

Figure 5–3 *Energy Savings: Incandescent Versus Fluorescent Light Bulbs*

erators sold in 1972 or earlier used less-than-optimum insulation, top-mounted freezer units, and automatic defrost, all of which are costly in terms of energy requirements. If all the households in the U.S. had the most efficient refrigerators available, the electrical savings could eliminate the need for about 12 large nuclear power plants.

Summary of Energy Efficiency Opportunities

Although improving energy efficiency does not have the glamour of space-based solar collectors or new fusion technologies, it is clearly the most promising of our energy options. It is estimated that total energy savings from the examples just described could, in the future, save the United States more than $100 billion per year.

NUCLEAR POWER: CAN THEY DO IT RIGHT THIS TIME?

Nuclear power is like Freddy the bogeyman in *Nightmare on Elm Street*—it just will not die or go away. Can nuclear power generation make a comeback? Oil and gas, as I have shown, are sure to become more expensive, and the greenhouse effect may preclude turning to coal. This leaves nuclear, solar, and some other exotic sources for power generation. One point is beyond argument: A dependable and economical supply of electricity is mandatory if the United States is to maintain a competitive position in the world economy.

The dangers and unanswered questions of nuclear power bother a lot of us. It is definitely not our energy source of choice, but it may very well be our energy source of necessity.

If nuclear power generation is in your future, what should you know about it? What is the difference between fission and fusion, for instance, and what in the world is a breeder reactor? Most important of all, can their designers make nuclear reactors safer? What are electrical utility companies going to do with the nuclear waste? In short, can the nuclear industry do it right this time?

Where Do We Stand Today?

The United States now gets 20 percent of its electrical power from 108 reactors—14 more are under construction at this writing. Some of the other industrial countries rely much more heavily on nuclear power generation than we do: France, 70 percent; Belgium, 66 percent; South Korea, 53 percent.

For a variety of reasons—psychological, technical, economic, and political—the early promise of the nuclear industry has not been met. No U.S. utility has ordered a reactor since 1978. Even before the Three Mile Island and Chernobyl failures, there were problems involving huge construction cost overruns and no acceptable solution to the nuclear waste worry. The combination of public fears and economic pressures has limited the growth of the industry. But the technology of nuclear power generation has changed, and maybe some lessons have been learned.

Fission and Fusion: What Is the Difference?

Both fission and fusion are nuclear reactions, in that they change the structure of an atomic nucleus. *Fission* involves the splitting of nuclei of heavy elements such as uranium into smaller parts by striking them with a free neutron. Fission is the system used in the atomic bombs, and it is the basic system used today in nuclear power plants. Uranium is the fuel used because it "splinters" readily, releasing two or more neutrons, which in turn strike and splinter other uranium nuclei in a chain reaction. The result of the chain reaction is the release of energy—suddenly in the case of an atomic bomb and gradually under control in a power plant.

Fusion involves the combining of light nuclei such as hydrogen. The nuclei of the hydrogen atoms are joined together, or fused, at an extraordinarily high temperature, to form a single, heavy helium nucleus, ejecting high-speed neutrons in the process. The atoms resulting from the joining weigh slightly less than the ones that fueled the process and it is this difference in mass that has been converted to energy (remember $E = mc^2$).

Fusion is the process that makes the Sun and the stars burn and powers the hydrogen or thermonuclear bomb. Scientists are working hard to adapt the fusion process to nuclear energy production. The problem is achieving the high heat required to initiate the fusion process. So far we have only been able to achieve this high heat in the thermonuclear bomb, which uses fission of an atomic bomb as the trigger.

Fusion has a lot to offer. Once perfected—and there are a lot of technical problems still to be solved before it is—fusion could generate energy equivalent to 300 gallons of gasoline from a gallon of sea water—and there is a lot of sea water. In addition to having an almost inexhaustible fuel supply, fusion produces relatively minimal waste.

Even the experts, however, are not optimistic about fusion power in the near future. Stephen O. Dean, president of Fusion Power Associates, has been quoted as saying that fusion in the twenty-first century would probably not be economically competitive with other energy sources. He is of the opinion, however, that environmental factors will spark the birth of the fusion industry in the twenty-first century. Coal- and oil-fired power plants make acid rain and warm the Earth's surface through the greenhouse effect, and conventional reactors make radioactive waste. According to Dean, environmental factors will eventually make fusion the winner. The scientific community is not without its skeptics, though, and some say that research on controlled fusion is a waste of resources. And then came the electrifying announcement of cold fusion.

Cold Fusion

B. Stanley Pons, professor of chemistry at the University of Utah, and his colleague, Martin Fleischmann of the University of Southampton in England, touched off a furor by asserting in March of 1989 in Salt Lake City that they had achieved nuclear fusion in a jar of water at room temperature. They claimed that this so-called cold fusion occurred when an electric current was passed through a palladium electrode immersed in water that had been enriched with deuterium, an isotope of hydrogen. The Utah team said that the palladium absorbs deuterium atoms, which are forced to fuse together, generating heat and neutrons.

The stakes could not have been higher. If true and if other scientists could have duplicated the Pons and Fleischmann experiment, the world might hope for an unlimited source of cheap energy. From the onset, however, physicists expressed profound skepticism of the claims made by Drs. Pons and Fleischmann. Despite an intensive cold-fusion research effort involving more than 1000 scientists and an estimated $1 million a day cost, little support for the Utah researchers' claim was forthcoming. At this writing the issue is still in some doubt, but the results are now largely discounted.

In the meantime we are dependent on fission, so let's see what the prospects are there.

Fission Reactor Fundamentals

A nuclear power plant is a complex assembly of steam generators, turbines, cooling systems, monitoring and inspection equipment, and power transmission lines. The heart of the complex is the reactor vessel, which contains a central core. The core contains a group of control rods made of a neutron-absorbing material, usually boron or cadmium, that are moved into or out of the core to regulate the rate of fission, and some rods of energy-emitting fuel. The large reactor vessel is the tower you see as you pass a nuclear plant. It is a simple structure, and as long as a constant source of cooling is available it is not dangerous. Dissipating the high residual heat in the core is the biggest potential problem with nuclear power. The basic differences between the nuclear power plants now in operation and proposed new designs lie in how they efficiently dissipate heat.

In the two reactor designs now in use in the United States, the heat of the chain reaction within the fuel rods is dissipated by ordinary water. In the *pressurized water reactors*, steam generators convert the hot water from the reactor vessel to steam, which then drives a turbine and produces electricity. In the *boiling water reactors*, the steam is generated directly inside the vessel.

How safe are current water-cooled reactors? A quick review of what happened at Pennsylvania's Three Mile Island in March of 1979 can help answer that question. TMI had a system of four barriers to prevent radiation from reaching the world outside and three of these failed for technical reasons or through human error. In an unlikely combination of equipment failure and control room errors, the fuel core was allowed to overheat. When a malfunction in the cooling system occurred, the TMI operators misread their instruments and instead of flooding the fuel core they turned off the backup cooling system. They did exactly the opposite of what was necessary. For 40 minutes the uranium core was partly or almost entirely uncovered by coolant. More than half the fuel elements melted into a blob at the bottom of the reactor vessel. The fourth and last of the defense barriers, a thick, steel-reinforced containment building, did prevent nearly all of the radiation from escaping. While core damage was major, the release of damaging radiation was minimal—one millionth of Chernobyl's. Nevertheless the near-disaster was too close for comfort, and while no one was injured in the TMI accident, it was almost the knockout blow for the U.S. nuclear reactor industry.

Can Nuclear Reactors Be Made Safe?

Instead of using multiple barriers and emergency pump systems, new reactor designs now under consideration abandon water as a coolant in favor of liquid sodium, which has a higher boiling point and a superior ability to absorb heat surges. Unlike a water reactor's coolant, which must be kept at high pressure to prevent the water from turning to steam, liquid sodium needs no pressurization. If pumps break down, the liquid sodium itself can prevent overheating indefinitely.

Because liquid sodium can explode when exposed to air or water, other precautions must be taken. The reactor vessel would have to be double-walled, with the space between filled with nitrogen as an added safety measure.

Another approach to a safer reactor was developed in Sweden and is based on the natural tendency of liquids of different densities to separate. In this design, a water-cooled reactor is surrounded by a large pool of cold, heavier water containing boron. If an accident occurs and the regular water coolant is suddenly lost, the heavier borated water rushes in automatically, keeping the core safely cooled. Borated water halts nuclear reactions by absorbing neutrons and thus acts as an emergency barrier for at least a week, while the operators effect emergency repairs to the water system. Figure 5-4 shows the various reactor designs.

This and other new reactor designs represent major improvements in safety, but there remains the major concern about nuclear waste. Our nuclear industry accumulates 1700 tons of highly radioactive "spent fuel" each year. As I discussed in the last chapter, underground storage is the official solution to the problem and Yucca Mountain in southern Nevada is the place. How safe will this be? Energy Department estimates show that only six inches of rain fall per year in this arid region, and only about 0.02 percent of this water would reach the buried fuel canisters. If radioactivity leaked out of the repository, it would take at least 9000 years for the small amount of contaminated water to transport to the water table 1000 feet below the repository. The underground rock, which is composed of compacted volcanic ash, would slow the movement of radioactive particles. The Department of Energy tells us that nuclear fuel loses its potency in 10,000 years.

This would all be reassuring were it not for DOE's track record. DOE experts at the Savannah River nuclear weapons plant once predicted that plutonium dumped in pits would take at least 1 million years to reach the water table. Less than 20 years later, plutonium at dangerous levels was detected in on-site groundwater—a pretty fair margin of error.

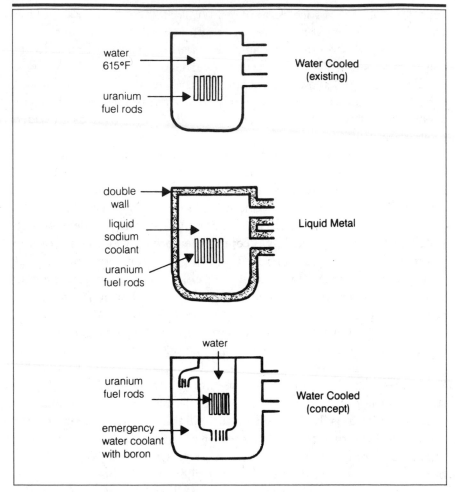

Figure 5–4 *Various Nuclear Reactor Designs*

Despite new and safer reactor designs and DOE assurances that the waste control problem is solved, nuclear fission reaction systems face serious public opposition. As if that were not enough, there is one other problem that may hinder a fission future: the fuel supply of uranium is limited. Without breeder reactors to generate fuel, fission power generation would have a short life span.

What Is a Breeder Reactor?

A nuclear breeder reactor has two purposes: it generates electrical energy and at the same time it produces fuel for fission reactors. Through the magic of the breeder process, a load of fuel generates both electricity and a new load of fuel—enough new fuel to feed itself and a second reactor. A breeder reactor extracts 60 to 80 percent of the fuel's total energy, compared with 1 percent efficiency of a conventional fission reactor.

Both breeders and conventional fission reactors rely on splitting atoms (fission) to generate heat, which ultimately creates steam to drive electric generators. The breeder, however, also uses the fission process to convert uranium 238 into plutonium. From each fission of uranium in the core, more than one plutonium atom is manufactured. In other words, more fuel is created than is consumed.

Despite the advantages of breeder reactors, not a single commercial breeder is in operation in this country today, and it is unlikely one will be operational in this century. The United States does have one demonstration breeder reactor at Clinch River in Tennessee. It has been involved in political controversy since construction was authorized in 1972. There are three main reasons for the controversy surrounding breeder reactors: safety, costs, and the danger of weapons proliferation.

The same safety concerns that give us pause in the case of fission reactors are quadrupled where breeder reactors are concerned. Breeders create plutonium, and this material is considerably more dangerous and longer lived than uranium. Some researchers consider it the most poisonous material in the world. If even small amounts of this material were to escape into the atmosphere, it would be a major catastrophe. The nuclear power experts tell us that breeder reactors are "just as safe as conventional fission reactors," but to many of us, that's not too comforting.

It has been so expensive to develop commercial breeder technology that Congress has balked more than once at the price tag. Even France, which is the leader in this technology, has recently cut back on its breeder program. The price of uranium would have to skyrocket before the breeder's fuel cloning ability would be cost-effective.

Now consider weapons proliferation. Spent fuel from a breeder reactor must be reprocessed to recover the plutonium created in its operation. The reprocessing plant is the one stage in the fuel cycle where material directly usable in weapons manufacture would be available. The storage and transportation of this "weapons-grade material" is a cause of concern. The fear is that this material could be stolen and used by terrorists to build a nuclear bomb.

Are the fears regarding accident or theft of material unreasonable? Justified or not, public fear will play a major role in the future of the nuclear power industry, and overcoming this phobia will not be easy. Remember the question raised at the beginning of the discussion about the ability of the nuclear power industry and the related government agencies to do it right this time? The answer is a qualified "maybe" and this only if a technically informed public keeps a close and skeptical eye on them. Most of us would be happier with a more benign source of energy such as the Sun but is that just a dream?

SOLAR POWER: ADVANTAGES AND DISADVANTAGES

Solar power techniques and cost effectiveness have failed to achieve the sunny forecasts of the 1970s. If you understand why this occurred, you can appraise solar power potential more realistically for the future. Most applications of solar energy today involve its conversion to heat or to electricity. This section briefly discusses the prospects for each.

Photovoltaic Technology

The photovoltaic effect occurs when light energy strikes two dissimilar materials and induces the generation of electromotive force. In that sense, a photovoltaic or solar cell is a solar battery. Photons of light energy striking a thin wafer, usually silicon, free electrons that move toward the positive pole while the holes from which the electrons were dislodged move toward the negative pole. An electric current results.

The cost of solar power depends on three factors. The first is the efficiency of the solar cells—the percentage of light striking the cells that is converted to electricity. The more efficiently this is done, the fewer the cells required to produce the same amount of electricity. The second factor is the cost to produce the cells. Finally, there is the overall capital expenditure required to install, operate, and maintain a solar power facility.

These factors sometimes work against each other. In general, the more efficient cells cost the most. Because of trade-offs, it could be more cost effective to use a lot of cheaper, lower-efficiency cells for some specific uses.

The inexpensive cells used on pocket calculators have efficiencies of 3 percent or less. High-cost cells used on our space projects have efficiencies of up to 20 percent. Recently photovoltaic technology passed an important milestone with the development of the first solar cell to surpass 30 percent efficiency.

Such a breakthrough was accomplished by stacking two cells. The upper cell of gallium arsenide reacts to blue light, whereas the lower cell made of silicon responds to the red portion of the spectrum. The stacked cells are mounted under an array of flat, plastic lenses that focus and concentrate the sunlight up to 500 times. Figure 5-5 illustrates this concept.

Because of advancements in solar cell technology and reduced manufacturing cost, a few solar power plants were built to sell electricity to utilities on a peak-demand basis. Within a decade or two, according to experts, solar plants could make a significant contribution to the nation's electricity needs.

How significant? Dan E. Arvizu, director of the photovoltaics program at the Sandia National Laboratories in Albuquerque, has said that despite the disappointments of the past decade, the recent progress indicates that photovoltaic energy could provide as much as 1 percent of America's electricity in the early years of the twenty-first century. The present United States electrical power generating capacity is about 600 billion watts. One percent then would be 6 billion watts, enough to supply the needs of 3 million people.

Cost is key to a solar future. When the overall cost came down to 12 cents per kilowatt hour, utilities turned to solar energy for such specialized uses as meeting peak power needs. When the overall cost has been brought down to 6 cents per kilowatt hour, solar power will come into widespread use in the generation of electricity.

Some scientists have proposed placing large arrays of solar cells in stationary orbit around the Earth and using microwave transmitters to convey the power to land-based receivers. Because sunlight is more intense outside the atmosphere and also not vulnerable to weather or

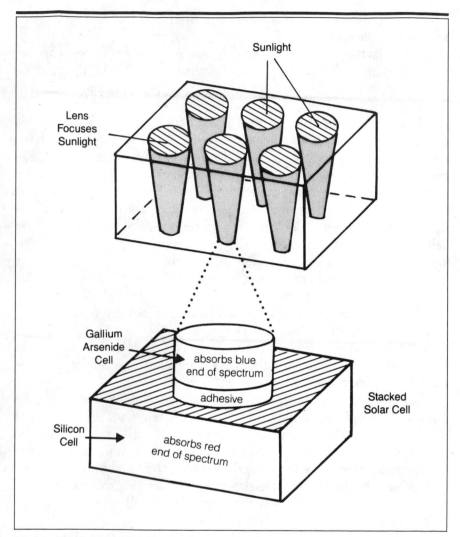

Figure 5–5 *Improved Photovoltaic Technology*

cloud cover, it is theoretically possible to increase production per unit area as much as 60 times. Even ardent advocates of this system, however, know that it is not likely to come to pass in the near future.

Conversion to Heat

Sunlight can be absorbed in collectors mounted on rooftops or open space to provide hot water or space heat. A typical flat-plate collector consists of a rectangular box, a blackened insulated bottom, a copper absorber plate, and a cover made of transparent glass or plastic. Overall, about 90 percent of all flat-plate collectors used in the U.S. today heat water. Solar water heating works fine but in most areas of the country it is not cost competitive with natural gas.

In some regions of the world, however, solar water heating is competitive and cost effective. Israel, Japan, California, and the southwestern portions of the United States are areas where the use of solar water heating is growing. Almost 60 percent of Israel's households now have solar-heated water. Some 11 percent of Japanese houses use solar systems, usually for heating water. The Japanese government expects that 8 million buildings will be equipped with solar systems by 1990.

Solar space heating systems involve storage problems and much larger amounts of heat. For these reasons, solar technology has not become competitive with other forms of space heating. It's hard to be optimistic about solar power for water or space heating. Formidable technical and financial constraints will prevent solar space and water heating from becoming a major player in the energy game, except for certain areas of the country where special conditions prevail. The real future for solar energy lies in its efficient conversion to electricity. Solar power will start to make sense (and cents) when the costs per kilowatt hour are competitive with other sources of electricity, and that in turn depends on the development of photovoltaic technology.

Although solar energy is a major hope for the future, it isn't the only one we have. Take a look now at some of the more imaginative energy sources for our future.

EXOTIC ENERGY SOURCES FOR TOMORROW

Hydrogen

It has long been known that hydrogen could be an almost perfect energy source. Nonpolluting and theoretically obtainable from ordinary water, hydrogen burns more efficiently than hydrocarbon fuels, and hydrogen combustion yields no toxic substances, only heat and water. With all of these possible pluses, why are we not planning a hydrogen future? There are, it seems, tough technical and cost problems to solve first.

What are the major disadvantages of hydrogen as an alternative energy source today? Cost of production is the major problem, and until some technical breakthroughs occur, costs will remain the limiting factor for some time to come.

Two other problems with hydrogen are storage and transportation. Hydrogen has a low energy density on a volume basis. This prevents storage of sufficient quantities onboard an automobile, for instance, to provide long-range operation comparable to the use of gasoline.

By most techniques, hydrogen is produced as a gas. This gas can be compressed and stored, but the volume and pressure requirements make it feasible only on a large scale. Cryogenic (very low temperature) technologies have made it possible to store or transport liquid hydrogen on a smaller scale. Pipelines, tanks, and trucks have all been developed to handle liquid hydrogen, but the expenses, of course, are high.

Despite drawbacks, the potential for hydrogen as a fuel is such that intensive research is continuing. The Soviet Union has flown a commercial airliner powered in part by liquid hydrogen on an experimental basis.

Commercial hydrogen is made by a process called *steam reforming*, but this process is expensive and virtually no industrial hydrogen is used for fuel because of its high price. Some investigators believe that the cost problem can be solved by using new technologies, such as photolysis and electrolysis for extracting hydrogen from water. *Photolysis* is a chemical reaction that uses light energy as a driving force. *Electrolysis* uses electrical energy for the extraction.

Because these new technologies promise lower cost, some hydrogen proponents believe that a major switchover to hydrogen is inevitable. They foresee hydrogen beginning a significant penetration of international markets in 25 to 30 years.

Wind Farms

Wind power machines range from simple rural water-pumping devices to large, modern electricity-generating turbines with 300-foot blades. The United States has been the pioneer in the development of these large wind-driven turbines and their use on wind farms.

At one time California had over 1000 wind machines in operation with a capacity of 60 megawatts located at a dozen wind farms. The California goal was for the state to have 4000 megawatts available from expanded wind farms by the year 2000. When the cost of oil declined and tax incentives were removed, wind machines became less attractive as investments. The California goal subsequently was revised downward.

Wind farms are not without their problems. They are restricted to specific areas where winds are reliable and they take up a lot of real estate. They are noisy enough to be annoying to neighbors, and they interfere with television reception. Community opposition is therefore a major hurdle to the expansion of wind farms.

Biogas Digesters

Neither exotic nor glamorous, the use of organic waste from plants and animals could nonetheless provide a significant source of energy, particularly in rural areas. At its simplest, a biogas digester consists of an airtight pit or container lined with brick or steel. Waste matter put into this container is fermented anaerobically (without oxygen) into a methane gas that can be used for cooking, lighting, or electrical generation.

China has been the world leader in the use of biogas digestion. The Chinese have over 7 million biogas digesters in operation—enough to meet the energy needs of 35 million people. Altogether, Chinese biogas digesters produce the energy equivalent of 22 million tons of hard coal.

Opportunities for generating biogas from animal wastes in the United States are limited to large dairy farms or feedlots. If all the wastes from the more than 15 million head of cattle in feedlots in the United States were converted to biogas, enough energy to heat a million homes could be produced annually.

Manure Fueled Power Plant

A pilot power plant in Southern California fueled by cow manure is now in operation, generating enough electricity to supply 20,000 homes. It is the first commercial power plant in the United States that burns only cattle chips for fuel. Conveyor belts feed 40 tons of manure an hour into the plant, where it is put into special furnaces. The heat produces more than 150,000 pounds of steam hourly to drive a turbine and electric generator.

Ocean Thermal Energy Conversion

Relatively small differences in water temperature at different depths can be converted into useful energy. The Earth's oceans absorb vast amounts of sunlight, most of which is radiated back into the atmosphere. A small fraction of this heat can be used in those areas of the ocean where the temperature difference between the warmer water on top and the cooler water below is at least 60 degrees Fahrenheit (15 degrees Centigrade).

Heat flows spontaneously from a hot region to a cold region. By channeling flow through a heat-driven engine, it is possible to redirect a fraction of the heat energy to useful work. The technology required to harness these temperature gradients exists. The most serious problems involve the mooring of large buoyant structures below water in the deep ocean, the stabilization of long undersea intake lines, and the transmission of the electricity from remote ocean plants.

Both the United States and Japan have spent more than $100 million on researching this technology, but so far little progress has been made. In addition to major technical challenges, environmental problems resulting from large-scale use of ocean thermal gradient plants are one of the constraints.

Magma Energy

Scientists are currently at work on a project to see whether it is possible to generate electricity by tapping the immense reservoir of heat inside the Earth. A team from the federal government's Sandia National Laboratory has been working in California's Sequoia National Forest for the past year drilling deep into a volcanically active area. Their objective is to see whether, using current technologies, a well can be drilled to the edge of a hot upwelling of magma, or molten rock, that is thought to lie three to six miles below the surface.

If this experimental project is successful, it could be the first step toward unlocking the potential of a vast new energy source that would dwarf the nation's known supply of fossil fuels. It is, however, a long-range technology that would not become available until after the year 2000.

SUMMING UP

This chapter discussed only a few of the possible alternatives to our current combustible energy generation system. Although it's impossible to foretell the future, energy policy arguments will be part of an ongoing political and societal debate. This chapter and the previous chapter on environmental problems are clearly interrelated. In addition to environmental factors, economic and geopolitical issues will influence energy choices throughout the world. How can China, for instance, create the energy for its industrialization, without using its vast coal reserves, thus contributing to the greenhouse warming? These and other energy-related questions will be the subject of worldwide debates for years to come. The technologically literate among us will be able to contribute meaningfully to these future energy discussions.

KEY CONCEPTS

▶ We obtain most of our energy in the United States today from petroleum, coal, and natural gas. We expend this energy primarily on electrical generation and on transportation.

▶ Change to alternative sources of energy may be required soon for two reasons: (1) our dependency on foreign oil seriously damages our economy and this dependency is bound to get worse, and (2) possible future environmental effects of the burning of fossil fuel—specifically the greenhouse effect—may lead to limits being placed on the use of coal and petroleum.

▶ Improving energy efficiency offers a means of buying time for our eventual conversion to new energy sources.

▶ Nuclear power—both fission and fusion—will be required to meet future national energy needs, and safer reactor designs are technologically feasible now.

▶ When viewed realistically, cost is the key to a solar-powered future. When cost has been brought down to about 6 cents per kilowatt hour, solar power will come into widespread use in the generation of electricity.

▶ Exotic power sources such as hydrogen, wind farms, and other alternatives to fossil fuel are long-range technologies that offer little potential until well into the next century.

6

What's So Important About

SUPERCONDUCTIVITY?

WHAT IS SUPERCONDUCTIVITY and why has it captured so much media attention? Is this phenomenon a key technology in this country's future, as some claim, or has the science been overblown and overhyped?

We've been promised many dazzling new technological advancements, including lightweight motors, wasteless power transmission lines, long-term energy storage systems, high-speed magnetically levitated trains, pollution-free electric cars, and hyperfast computers. In its power to transform our physical world, superconductivity technology has been compared to the invention of the light bulb and the transistor.

The media, caught up in superconductivity fever, overdid the hype. The excitement engendered by the discoveries and breakthroughs of a few years ago got somewhat out of hand, and major developments in such fields as transportation and electrical power generation and transmission may lie years in the future. For example, *Time*'s cover story (May 11, 1987) entitled "The Superconductivity Revolution" was largely absurd. To power the futuristic automobile shown on the cover of the magazine by means of a superconducting magnet would require an impossibly large magnet—as large as an eighteen-wheel truck.

This is not to say that there is not great promise in superconductivity or that the current hectic pace of research is not justified. It's just that we should all take a deep breath and try to take a more rational look at the phenomenon. One of the advantages of technological literacy is to be able to tell the hype from legitimate science and there is plenty of exciting legitimate science in superconductivity.

What happened to catapult a relatively obscure phenomenon, first discovered in 1911, to its present prominence? This prominence, by the way, has been marked by almost frantic research activity. At a 1988 meeting of the Materials Research Society, for example, some 260 papers on superconductivity were presented.

SUPERCONDUCTIVITY FUNDAMENTALS

Electricity is the flow of electrons. The material through which the electrons flow is called a *conductor*. Different materials vary in their ability to conduct electricity, but all materials offer some resistance to the flow of electrons. A percentage of the flow of electrical energy is thus lost (usually in the form of heat) because of resistance. For instance, copper we know is a good conductor as is aluminum, silver, or gold. Other materials such as wood, glass, or rubber are not good conductors. Superconductors are materials that conduct electricity with practically no resistance at all; none of the electrical energy is lost when flowing through a superconductor.

Electric current—the flow of charged particles through a conductor—can be likened to water flowing through a pipe. Voltage is the pressure pushing the water through, and conductivity can be thought of as the size of the pipe through which the water flows. The larger the diameter of the pipe, the less resistance to the flow and thus the greater ability to conduct.

Prior to 1911, there was no way to eliminate the resistance inherent in even the most efficient conductors. With the discovery of superconductivity at that time, a new type of conductor was developed.

The reasons for resistance in a material have to do with the atomic structure of the material. At ordinary temperatures, thermal vibrations of the material's atoms obstruct the flow of electrons. Superconducting occurs in some materials when these vibrations are suppressed at exceedingly cold temperatures.

An analogy of superconductivity is a playground swing. If you give it a hard push, it swings in an arc for some time, but no matter how hard you pushed to begin with, the arcs will become increasingly shorter until the swing comes to a stop. The friction of the air resisting the swing and the friction of the chains or ropes where they are connected to the structure as well as the pull of gravity bring the swing to a halt. Now imagine that the resistance or frictions have somehow been eliminated. The swing, once pushed, would swing at the same rate forever. So it is with a superconductor. In a normal conductor, the current will quickly diminish due to resistance. In a superconductor, the current can continue to flow for a long period of time because there is practically no resistance to stop it.

Although scientists have long known about the phenomenon of superconductivity, it remained a laboratory curiosity until 1987, because all the materials used had to be refrigerated down to a very cold 23 degrees Kelvin (K) or less before they would function as superconductors. (Kelvin is a measure of the degrees above absolute zero. Zero Kelvin is the same as −460 degrees Fahrenheit or −273 degrees Centigrade.) Table 6–1 shows the three main temperature scales in use today and their relationship.

Don't worry, you won't have to do any math. In this country we've all grown up with the Fahrenheit scale to the point where we have a feel for it. We know that a temperature around 70 degrees Fahrenheit (21 degrees Centigrade) is comfortable and that when it gets to 80 or 100 degrees Fahrenheit (26 or 38 degrees Centigrade) outside it is hot.

Table 6–1 Temperature Scales and Conversion Formulas

Action	Degrees Fahrenheit	Degrees Centigrade	Degrees Kelvin
Water Boils	212	100	373
Room Temperature	68	20	293
Water Freezes	32	0	273
Nitrogen Liquefies	−320	−196	77
Helium Liquefies	−451.84	−268.8	4.2
Absolute Zero	−459.4	−273	0

CONVERSION FORMULAS

Degrees Fahrenheit = (9/5 × degrees Celsius) + 32

Degrees Celsius = 5/9 (degrees Fahrenheit − 32)

Degrees Kelvin = Degrees Celsius − 273

Most other countries, however, use degrees Centigrade (Celsius), and when we finally convert to metrics, we will change over to the Centigrade scale. The Kelvin scale starts at *absolute zero*, the point at which molecular motion stops. Kelvin is the temperature scale used almost exclusively for scientific work. It is the scale I will use in the following discussion. Just remember that the higher the positive K value, the warmer the substance gets.

Prior to the recent breakthroughs in the technology, the only way to get a material cold enough to superconduct was to refrigerate it with liquid helium (4.2 Kelvin), and liquid helium is expensive and difficult to work with. The result was that superconductivity was confined to the laboratory and a few specialized uses. All this changed dramatically when superconductivity came in from the cold.

DISCOVERY OF HIGHER-TEMPERATURE SUPERCONDUCTORS

In January 1986, Karl Muller and George Bednorz, scientists with IBM in Zurich, Switzerland, discovered a ceramic material able to superconduct at 30 degrees Kelvin (raised later to 39 degrees Kelvin), significantly higher than any previously known substance. Superconductivity in a ceramic was unheard of until this time. This exciting discovery prompted many superconductivity researchers to experiment with similar types of ceramic materials. The discovery not only won the Nobel Prize for Drs. Muller and Bednorz, but it initiated the superconductivity race.

In February of 1987, Dr. Paul Chu and his team at the University of Houston hit upon an astonishing, totally new variety of superconductor that operated at much higher temperatures than most experts had thought possible: 98 Kelvin. A significant barrier had been broken. Nitrogen liquefies at 77 Kelvin, and liquid nitrogen is relatively inexpensive (less than 50 cents a liter versus several dollars a liter for helium) and can be transported in a thermos.

The 77 Kelvin temperature mark is still extremely cold, mind you (comparable to -196 C or -320.8 F), but an important benchmark nonetheless. The 77 Kelvin point has been compared to the sound barrier or the four-minute mile. Below this temperature, superconductivity remained a technical curiosity, but above 77 Kelvin there are almost no limitations on the material's practical application. The race to raise this critical temperature even higher is continuing. Scientists have set their sights on a room temperature superconductor that would not require any cooling. In the meantime, applications of the current superconductors are being tested in a number of areas.

POWER SYSTEMS

Superconductors have great potential for improving the efficiency of power generation, distribution, and consumption. *Power systems* here include all the equipment and methods used to generate electricity, distribute it to the consumer and the electrical appliances that use it. Figure 6–1 is a simplified sketch of the overall power system. Superconductivity has a role to play in each phase.

Power Generation

There are already several superconductive electrical generators in existence. These experimental generators were all made using the old low-temperature superconductors. The great advantage of superconductors is that they are capable of carrying large currents through small areas permitting the superconducting generator to be much smaller with the same power output.

Although these experimental generators did prove that superconductivity would work in generators, their costs were too high for practical use. The breakthrough to high-temperature superconductors may eventually change this picture, but as of this writing the new ceramics are too difficult to form and they cannot carry the required high currents. It will probably be a number of years before the new high-temperature superconducting materials have been developed to the point where they can be used in large-scale generators.

Power Storage

Power storage is an area where superconductivity does have some promise. There are currently few ways to store large amounts of electricity. The utilities must, therefore, be capable of generating the max-

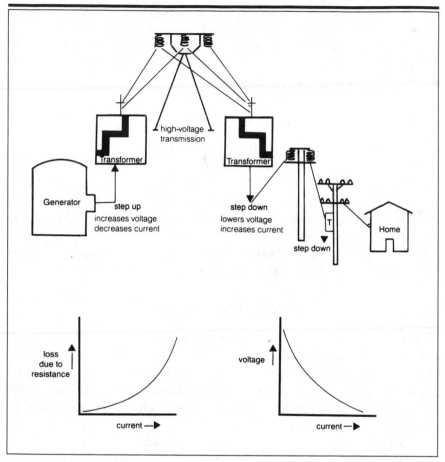

Figure 6–1 *Electrical Power Transmission System*

imum power to meet the peak needs of a community, which results in overcapacity during off hours. A large superconducting coil (in the form of a circular continuous cable placed underground) could stockpile electricity in the same way that a reservoir stores water. This allows generators to be run at a constant rate full-time and still meet the peaks and valleys of daily demand. Such load leveling is now accomplished by using excess electricity to pump water upstream and into a reservoir. The water can be used later to generate more hydroelectric power. Experts tell us that such pumped hydrostorage is more economical today than superconducting storage. In the future, with a reduced cost for cooling and materials, superconducting storage will become com-petitive.

Design and engineering studies are now underway for a system called Superconducting Magnetic Energy Storage (SMES), which might someday provide a means for the large-scale storage of electrical energy. Large superconducting coils, placed underground and refrigerated, would be capable of storing large amounts of electricity for long periods of time. A SMES system would be connected to a power grid and used to supply power during peak load times.

Power Distribution

Using superconductivity for power transmission lines is a possible area for a large-scale application. In current power distribution systems, losses due to resistance amount to 5 to 10 percent a year. Replacing conventional conductors with superconductors would greatly reduce or even eliminate this energy loss. There are, however, economic ob-stacles. In 1985, the United States consumed $175 billion worth of electricity, sustaining from $8 to $17.5 billion in transmission losses. Even if superconducting lines were able to save $10 billion a year, this figure has to be compared to the capital costs of installing such lines, which might be hundreds of billions.

Right now the major problem with superconducting power cables is that because they have to be refrigerated, they must be placed un-derground. Over long distances, overhead transmission wires are from 10 to 30 times cheaper to construct than underground cables.

Eventually, the benefits of superconductivity will be felt in power distribution systems. The short-term problems pointed out here do not really diminish the long-term potential. The electrical carrying ability of a superconductor is so high that just one cable could supply an entire city. Conventional resistive transmission lines sustain losses of about 1 percent per 100 miles. That is the reason there is no trans-continental power grid in the United States today. It is cheaper to build new power stations than it is to install long lines. The possibility of nitrogen-cooled superconductors suggests the feasibility of conti-nent-size grids. Power grids on such a large scale would permit the development of electrical resources in places that appear too remote today. The vast hydropower resources of northern Canada, for in-stance, could be developed and tapped. Indeed, transmission lines thousands of miles long, laid as transcontinental links or undersea cables, might join the world in a common power pool.

Current estimates, however, are that it will take at least five years before superconducting power lines become a practical reality.

Power Consumption

Home appliances such as dishwashers, washing machines, and refrigerators may someday use superconductive electric motors that use much less power than today's appliances. Other, smaller appliances might use superconductive circuitry and wiring to cut down on power consumption. In fact, superconductivity could have a major effect on the whole field of electronics.

SUPERCONDUCTOR ELECTRONICS

Applying superconductivity to the design of computers or any other devices that use transistors and integrated circuits—stereos and television sets, for example—is an exciting prospect to designers of electronics.

Transistors

Transistors are the electronic switches that replaced vacuum tubes (also electronic switches) in radios, televisions, and computers. You may remember when we removed television tubes and tested them at the local supermarket in order to keep the set working. Transistors (solid-state devices) made appliance maintenance a lot easier. Transistors are made from semiconductor material—a substance that conducts electricity better than an insulator but not as well as a conductor.

Silicon and germanium are the substances used in most transistors. When these materials are combined and altered with impurities such as arsenic or boron, they can be made to act just like tubes in an electronic circuit. That is, they can be made to pass current in one direction and not the other (in which case they are called *rectifiers*) or they can be made to increase the current in a circuit (*amplifiers*). Because the transistors were made of solid material, the new technology was called solid-state electronics.

Integrated Circuits

As important as transistors were to the miniaturization of electronic circuits, one more development was necessary to bring us into the age of supersmall electronics. Wiring a bunch of transistors together in a complicated electronic circuit limited technological advances and complexity. One wiring error or one break in a tiny wire and the device didn't work. Engineers knew that a way had to be found to replace the number of individual components in a complex circuit.

A new field of microelectronics came into being. Instead of individual transistors, a small piece of silicon could be etched microscopically to form what are now called *integrated circuits (ICs)*. Eventually these small chips of silicon became small enough to be called *microchips*.

That's the background. Superconductivity has the potential to significantly advance electronic technology. By eliminating heat and magnetic interference problems, superconducting ICs can be packed with the transistors and other elements closer together. This results in faster operating times and more compact devices.

Recent Developments

Conventional semiconductor transistors use a small voltage to act as the on/off switch necessary to the binary arithmetic of digital devices. (Binary systems are described in Chapter 3.) The amount of heat this produces has until now been a crucial obstacle to miniaturizing computer components. Then came *Josephson junctions*. Named for its developer, Nobel Prize-winning British physicist Brian Josephson, the junction is a speedy electronic switch that operates 1000 times faster than a transistor and uses 1000 times less energy.

A Josephson junction consists of two superconducting films separated by an insulating layer that prevents current from flowing between the films when the junction is turned off and lets it flow when the junction is turned on. A Josephson junction can operate in less than 2 picoseconds (a *picosecond* is one trillionth of a second). This capability has the potential to increase the speed of electronic instruments, computers, or communications systems. IBM spent over $300 million on Josephson junctions in a grand effort to achieve a superconducting computer. They abandoned their research efforts in 1983, but Josephson technology has recently been used in more narrowly targeted products.

MEDICAL APPLICATIONS OF SUPERCONDUCTIVITY

The one large-scale superconducting application currently on the commercial market is in *nuclear magnetic resonance (NMR) imaging*, which uses the magnetic properties of the protons of hydrogen atoms to generate detailed images of the human body's soft tissues. (See Chapter 7 for details of this marvelous machine.)

NMR machines cost between $1 and $2 million, and the use of helium to cool the superconducting magnet is a major contributor to its high annual operating cost. Liquid-nitrogen coolant for the newer high-temperature superconductors is, as I've said, much less expensive and could result in significant savings.

Because the Josephson junctions I described earlier are extremely sensitive detectors of magnetic fields, their use in measuring minute magnetic fields is useful in medical applications. *Superconducting quantum interference devices (SQUIDs)* use this sensitivity to measure magnetic fields a thousand times smaller than is possible with any nonsuperconducting device. The SQUID can be used as a supersensitive voltmeter, able to detect voltages as low as a billionth of a billionth of a volt.

An important application of SQUIDs is in a medical field known as *magnetoencephalography*. The SQUID is used to measure small magnetic fields produced by the firing of neurons in the human brain. Magnetoencephalography has two main advantages over its electrical counterpart, electroencephalography (EEG). It is completely noninvasive—that is, no surgery or injections are required—and it can locate the source of specific signals in the brain. The SQUID method can pinpoint brain cells that cause epilepsy and may some day permit the early diagnoses of Alzheimer's and Parkinson's diseases.

One of the strangest manifestations of superconductivity is levitation, in which a magnet hovers in midair above a piece of superconducting material (you'll read about this application in the next section). In 1988, scientists discovered an equally startling phenomenon, which they call the *suspension effect*. A chip made of superconducting material can be made to hang below a magnet as though suspended by an invisible string. As long as the chip stays cold enough, it remains suspended.

Using this suspension effect, a surgeon might be able to perform delicate brain surgery without opening the skull. A flexible strand of material with a magnetic tip can be threaded through the blood vessels leading to the brain and manipulated there by the surgeon.

As more superconductivity technologies are developed in the laboratory and make their way to the health care field, we can expect more applications. Other applications that might be ready in the near future include

- Sensors that can detect disturbances in the Earth's magnetic field caused by a submarine operating deep in the ocean.

- Antennas that can receive signals at frequencies far higher than possible today for use in communication and space exploration.

LEVITATING TRAINS

One of the most dramatic applications envisioned for superconductors is the magnetically levitated (MAGLEV) train. No article on superconductivity in the popular press has failed to show an imagined superfast train floating on a cushion of magnetic field supplied by superconducting magnets. The concept is not as fanciful as it sounds. Prototype levitating trains using the conventional superconductors made from niobium-titanium and cooled by liquid helium have been built and tested by both Japan and West Germany. The Japanese MAGLEV train has set a world rail speed record of 321 miles per hour (517 kilometers per hour) on a 4.2 mile test track (actually not a track with rails but a guideway). The Japanese announced goal is an operational Tokyo-to-Osaka line sometime in the 1990s.

Figure 6–2 shows how the train, with superconducting magnets placed along both bottom corners, rides in a U-shaped guideway that has copper coils on each of its three inside walls. Electromagnetic forces between the superconducting magnets and these coils propel, levitate, and guide the train.

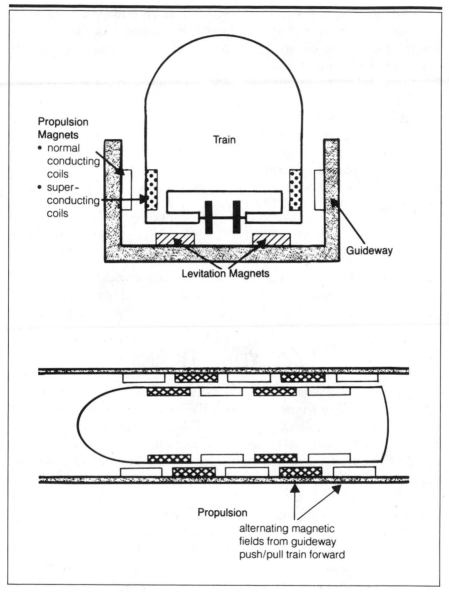

Figure 6–2 *Magnetically Levitated Train*

Each magnet on the train is attracted by a guideway coil of opposite polarity immediately ahead of it, and repelled by a coil of the same polarity immediately behind it. Rapidly alternating the polarity of the guideway coils propels (pulls and pushes) the train forward. The frequency of the polarity reversals governs the speed.

At about 20 miles per hour, the superconducting magnets on the train induce enough current in the guideway coils to produce a magnetic field of the same polarity as the superconducting magnets. The resulting repulsive force raises the train about 4 inches above the guideway.

Steering along the guideway is achieved in the same way. If the train deviates toward one side of the guideway, the superconducting magnets create currents in the coils mounted in the guideway walls. Polarity in the coils is controlled to repel from one side and attract from the other. The result is to keep the train in the center of the guideway.

Fast and quiet, the superconducting MAGLEV train seems ideal for certain U.S. transportation needs. The Boston-Washington corridor, for instance, might be heavily traveled enough to justify the development cost of a levitating train system. A closer look at the cost figures, however, raises doubts. The major expense in any levitating train system would not be in the superconducting magnets, which represent a relatively small part of the total costs, but rather in the guideway development. In short, the technology has been proven, but constructing a transportation system based on this technology may not be economically feasible at this time.

Proponents of MAGLEV technology are willing to take the financial risk. In an effort to get a toe in the American market, a West German manufacturer of MAGLEV technology has offered to build a MAGLEV train system between the Las Vegas airport and the downtown area. The Las Vegas system will be a modernized version of West Berlin's M-Bahn, the world's first public MAGLEV system, which went into operation in late 1988.

The West German company has offered to set up similar projects in other areas of the United States, but they face competition. At this writing, a British MAGLEV project is being considered for a transit route in Atlantic City. This same technology is also under consideration for a commuter line running to and from the city to New York's La Guardia Airport. All of these promoters want to have a running

MAGLEV system in place in the United States to demonstrate that it can be economical.

SHIPS, CARS, AND SUPERCONDUCTIVITY

The U.S. Navy thinks that both surface and submarine vessels could be running on superconductor power by the mid-1990s. The proposed ship use is not as spectacular as the levitating train but nonetheless has potential. Superconducting magnets can be used to make an extraordinarily compact electric motor. The Navy is planning to test a 40,000-horsepower superconducting motor measuring about 6 feet in diameter and weighing less than 80,000 pounds. A conventional electric motor producing the same torque would measure 16.5 feet in diameter and weigh nearly 300,000 pounds. Because of their relatively small size, superconducting motors may someday be used in submarines.

Cars, as I pointed out earlier, are not a good potential use for superconductors. A superconducting motor weighing 100 pounds could generate about 130 horsepower but only for a single second. Despite the technical obstacles, a Ford Motor Company research team is investigating the feasibility of electric cars using superconductivity. If it is possible to develop practical high-temperature superconductors in the future that do not require bulky cooling systems, the electric car may be revived. Now the imagination turns to the largest of the proposed large-scale uses for superconductivity.

WHAT IS A SUPERCONDUCTING SUPERCOLLIDER?

You now know what superconducting is, but what in the world is a collider (let alone a supercollider)? To learn the answer to that question, take a quick look at the rarefied world of high-energy physics.

High-energy physics, which is also known as *particle physics*, is one of the basic sciences. Elementary particles are the basic components of all matter. The atomic theory of matter says that the enormous variety of substances that surround us are made up of only a few different chemical elements, each of which has its own type of atom.

All of the known atoms have been classified and organized into a chart, the periodic table. The periodic table groups the similar types of atoms such as metals and gases and identifies each element by a symbol. Figure 6–3 shows a simplified diagram of an atom. The number of protons in the nucleus determines what material (element) it is.

Scientists, in their search for a better understanding of how matter and energy interact and transform, need to gain a knowledge of the forces that control elementary particles. To do this they use an apparatus called a *particle accelerator*. This device accelerates electrically charged particles (electrons and protons) to high speeds using magnetic fields. The particles are pulled and pushed by the magnets up to speeds approaching the speed of light when they are smashed into other elements or materials. The results give scientists information on how these high-velocity particles transform into other particles. Accelerators that accomplish this mission are called *colliders* (or "atom smashers" in the popular press).

The colliders enable the physicist to look inside the atom. The method of doing this has been called "the Swiss watch" technique. If you want to find out what's inside a Swiss watch, hit it with a hammer or bang two of them together and see what comes out. The best way to generate the forces required to propel the particles is using super-

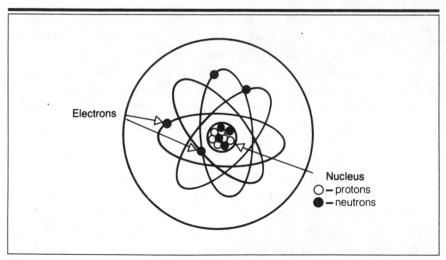

Figure 6–3 *Simplified Diagram of an Atom—The Number of Protons in the Nucleus Determines What Material (Element) It Is*

conducting magnets. The largest collider in the United States is called the Tevetron, located at Fermilab in Illinois. The Tevetron is a ring 4 miles in circumference containing 1000 superconducting magnets. It has been successful from the point of view of the scientists, but it merely whetted their appetite. Now they want a really big collider built—a supercollider.

In effect, what the physicists want is a bigger hammer.

Big is the right word. If and when it is built, the Superconducting SuperCollider (SSC) will be mankind's largest device ever, and, at $4.4 billion, one of the most costly. At a site in Texas, 25 miles south of Dallas in Ellis County, the SSC will consist of a racetrack-shaped tunnel 53 miles in circumference, 10 feet in diameter, and buried 150 feet underground (see Figure 6–4).

Inside the tunnel, two rings of superconducting magnets—9000 of them—will be spaced along the particle beam pipes to focus, propel, and guide two beams of protons traveling in opposite directions. The protons will race around the track, gaining momentum with each circuit until they are traveling near the speed of light. At special chambers called *interaction halls*, the protons will cross over and collide with an energy 20 times greater than has ever been achieved before. They are really going to smash those Swiss watches this time, and it will be a titanic collision.

One way to envision the size and cost of the SSC is to imagine an auto racetrack 20 times larger than the Indianapolis Speedway, its entire length covered with bumper-to-bumper top-of-the-line Mercedes. The Mercedes represent the 10,000 superconducting magnets set end to end and, in fact, cost about the same per foot.

Why this shape and size? Like many things, the SSC represents a compromise. The 53-mile circumference is a trade-off between size and magnet design. Original proposals for the SSC had the circumference from 50 to 100 miles. The larger circumference would have saved money on the magnets, but the site problem would have been exacerbated, and a larger tunnel would have increased the cost estimates.

With its $4.4 billion price tag, the SSC is not without its critics and doubters. The SSC is a so-called Big Ticket science project comparable to the space station ($23 billion) or the human genome project ($3 billion) discussed in Chapters 1 and 2. Congress may decide not to fund all of these programs in today's budget-conscious period.

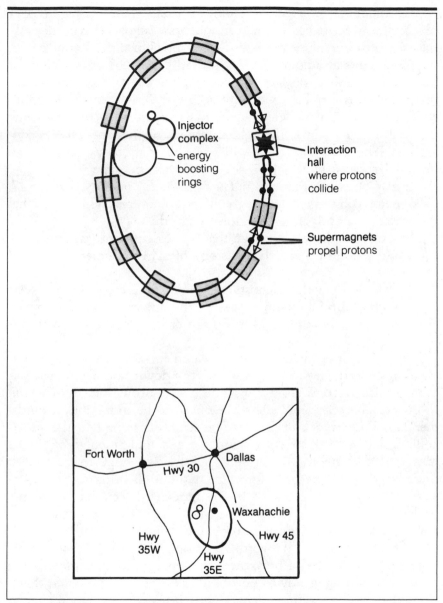

Figure 6–4 *Proposed Superconducter Supercollider*

Critics of the supercollider point out that studies in high-energy physics are unrelated to any conceivable practical use. What, they ask, have the physicists done for us since they developed the atomic bomb?

Opponents also point out that the SSC isn't the only option for keeping high-energy physics alive and well in the United States. Another option is for this country to accept the invitation to buy into the Large Hadron Collider, a somewhat similar instrument being planned at Geneva by CERN, the European high-energy physics consortium. The CERN machine will cost half as much as the U.S. supercollider and may be able to achieve almost the same results.

It may be, as an editorial in *The New York Times* (November 17, 1988) pointed out, that the supercollider epitomizes the luxuries that America is currently in no position to afford.

Proponents of the SSC claim that it is essential to understanding the world around us, and that it is one of man's premier intellectual challenges.

Specifically what the scientists are looking for are answers to several basic and profound questions: How did the universe come into being? Why do objects have mass? How do the basic forces that govern our universe behave?

Physicists have carried on their search for a fundamental understanding of matter by moving from the top downward, from materials to molecules to atoms. Inside the atoms, you go from nucleus to proton to quark. The scientists believe the quark to be indivisible, a fundamental building block of nature. The scientists hope that the supercollider will help them understand a *grand unification theory,* what they call "the theory of everything." Physicists hope that a unified understanding of the basic forces of nature will bring technological rewards of the kind spawned in the nineteenth century, when the understanding of the relations between electricity and magnetism paved the way for modern electronic technology.

If so, the SSC will be worth the money. In the past, every time pure science achieved a breakthrough, a major new technology has developed, adding greatly to our GNP and improving our life styles. True, the stakes are getting higher, but that may not be reason enough to quit the game (especially if the jackpot is as grand as this one).

For many years the Defense Department has trotted out the old ploy "The Russians Are Coming" to convince Congress to fund this or that weapons system. We are now seeing a variation on this ploy,

"The Japanese Are Coming," to talk Congress into funding large research projects of one kind or another. It's a persuasive technique and for that reason, if for no other, we are going to have a superconducting supercollider in our future.

The SSC is expected to be completed in the mid-1990s if current technology is used. If the newer higher temperature superconductors are to be used, the completion date will slip to the year 2000.

THE FUTURE OF SUPERCONDUCTIVITY

At this writing, the record high temperature is 125 Kelvin (-234 degrees Fahrenheit and -148 degrees Centigrade), but the scientific community is hard at work trying to push the critical temperature upward. The steady pace of achievement over the past two years adds to the confidence on the part of many researchers that large classes of superconducting compounds await discovery and that even room-temperature superconductors will eventually be possible.

In addition to the temperature problem, the producibility of new materials is an issue in the development of practical uses for the superconductivity phenomenon. In the summer of 1988, Stanford University scientists developed tiny superconducting fibers that can carry fairly large electrical currents, a development that is considered an important step in the commercialization of new superconducting materials.

Making wires of the new superconductors is necessary if the materials are ever to be used in most of the applications discussed in this chapter. So far, researchers have been able to get high currents only when they deposit thin films of superconducting materials on other substances. Wires made out of the superconducting material itself have not been able to carry large currents.

This inability to carry high currents has turned out to be the major stumbling block to achieving the dramatic potential of high-temperature superconductors. In mid-1989, researchers at the AT&T Bell Laboratories said that they had found that none of the new superconducting compounds could carry large electric currents, and that all of them cease to be superconductors when exposed to large magnetic fields.

The problem apparently relates to the behavior of lines of magnetic force that penetrate the superconducting material. These flux lines become distorted when the temperature is raised or the current is increased, thereby destroying the material's superconductivity.

At the present time, this electromagnetic phenomenon precludes all of the large-scale schemes for exploiting high-temperature superconductivity. The big question now is whether a technical solution to this problem can be found.

So what is the bottom line? What is the answer to the question posed in the title of this chapter? The importance of superconducting is its potential. After a series of dramatic findings over the past three years, superconductivity is now entering a new phase. The predictions of new applications are today far more modest than they once were. Experts now expect the first practical uses of the new technology to be in highly specialized microelectronics, primarily for the military.

Superconductivity can be viewed as a major technology that will play a preeminent role in our future high technology society, but the emphasis is on the future tense.

KEY CONCEPTS

▶ Superconductors are materials that conduct electricity with no resistance whatever.

▶ The phenomenon of superconductivity occurs only when a material is cooled to an extremely frigid temperature. The point at which superconductivity occurs varies among different materials.

▶ The race to raise the critical temperature at which superconductivity occurs is continuing and scientists have set their sights on a room temperature (293 Kelvin) superconductor that would not require any cooling.

▶ Superconductivity has great potential for improving the efficiency of power generation, distribution, and use.

▶ By eliminating heat and magnetic interference problems, superconductivity has the potential to advance electronic technology—specifically computer design—significantly.

▶ One of the most dramatic near-term applications envisioned for superconductors is the magnetically levitated (MAGLEV) train.

▶ All of the large-scale applications of high-temperature superconductivity depend on solving the electromagnetic problem, which at the present time threatens to block the dramatic potential applications.

▶ The proposed superconducting supercollider would be the most powerful instrument for probing the structure of matter ever built, but the question of funding the $4.4 billion machine raises issues of scientific priorities.

7

HIGH-TECHNOLOGY MEDICINE

AS WE EDGE toward the twenty-first century, exciting new technological developments are initiating a challenging new period in medical history. I can only skim the surface in one short chapter, but it is probably fair to surmise that at least one of the new technologies described in these pages will play a role in your health care, if in fact it hasn't already.

Physicians often disagree about what laypersons should know about health care. On the one hand, physicians do not want patients telling them what therapy to apply or what drug to administer—advice based on "something heard on a television program" or the "disease-of-the-month" from a current magazine. On the other hand, many physicians do find it helpful for their patients to know as much as possible and to be able to discuss alternatives and make decisions based on some knowledge of the technologies involved.

The state of health care in this country today is such that personal life-style assessment programs, home diagnostic kits, and even do-it-yourself treatments are playing a growing part in the overall medical picture. Laypersons, it seems, are going to have to assume more responsibility for their own health care.

But even if we didn't have to get involved, it is better to know as much about potential diagnostic and treatment technology as we can. The technological explosion is nowhere more evident than in medicine. For instance, recent advances in imaging technology, or machine vision, have enabled physicians to see inside the body without the trauma of exploratory surgery. As a result, more progress has been made in diagnostic medicine in the past 15 years than in the entire previous history of medicine.

Other advancements include surgical lasers; arthroscopy; micro-computers that can help diagnose an illness and then monitor and control patient care; and sensors that can measure internal substances, such as blood chemistry, without entering the body. Recent advance-ments in drug delivery technology as well as transplantation technology may have opened a new era in the treatment of brain disorders such as Parkinson's disease and Alzheimer's disease.

Do you know, for example, what differentiates PET scans, CAT scans, and magnetic resonance imaging? How do they work, and what are the results? This chapter summarizes the latest developments in the world of medicine.

COST FACTORS:
WHO WILL PAY FOR ALL THIS?

High-tech medicine, while offering great promise for improved medical care, is extraordinarily costly. Expect to pay far more for your future health care than you have in the past. In 1988, Americans spent $2135 on medical care for each man, woman, and child. The nation's medical bill in 1988 topped $540 billion—over 11 percent of the gross national product and almost twice the cost of national defense.

As you probably have noticed, the average health insurance bill jumped 20 percent to 40 percent in 1988. Overall medical costs have been growing at twice the rate of inflation. High-tech medicine is the major contributor to this inflation, although lawsuits and malpractice insurance also have contributed to these costs.

If you plan to have a heart transplant this year, the cost is $125,000. Liver transplants cost $160,000, and bone marrow transplants run $100,000 to $150,000. Genentech's bloodclot-dissolving drug, TPA, can help heart attack victims—at $2200 per dose. The most effective treatment in prolonging the lives of AIDS patients is a drug called AZT, but it costs each patient an average of $6500 per year.

Even if you stay healthy, the rising cost of medical care is going to affect you. Your health insurance will rise, as I mentioned, but in addition there will be hidden health costs. For instance, the price tag of a typical new American car now includes about $360 in employee health care benefits. In contrast, the health care factor in the markup for an average Japanese car is only about $100.

Table 7-1 shows how the national health bill has risen already and projects how high it is going to get. Why all this emphasis on costs in a chapter on medical technology? Because, in medicine, costs and technology are so closely intertwined that it is not possible to discuss one without the other. In fact, the cost of continued scientific break-throughs is beginning to outstrip the ability of most of us to pay for them.

As you read about the many medical marvels in this chapter, keep your hand on your wallet. Case in point: Magnetic resonance imaging or MRI (described a bit further on) costs about $800 per session.

In addition to increased costs, health care has changed in several other important ways. People used to go to the doctor when they became ill or had an accident, and doctors diagnosed the problem and prescribed treatment. It's not that simple anymore. Consider the pop-ularity of home diagnostic kits and preventative medicine—which in-volves visiting a doctor when people feel well.

TABLE 7–1 America's Rising Medical Bill

Year	National Health Spending (billions)	Per Capita Health Spending
1965	$ 41.9	$ 205
1970	75.0	349
1975	132.7	590
1980	248.1	1054
1985	422.6	1721
1990	647.3	2511
2000	1529.0	5551

Source: U.S. Department of Commerce.

HOME DIAGNOSTIC KITS

Home diagnostic kits have become available in drugstores in the past few years and have won qualified approval from the medical community as legitimate and valuable adjuncts to our more traditional health care. The test kits described in Table 7–2 are the most widely used.

A number of other over-the-counter home tests are being developed at this time, including tests for strep throat, allergies, thyroid problems, and sexually transmitted diseases.

Table 7–2 Common Home Diagnostic Kits Listed in Descending Order of Popularity

TEST	PURPOSE
Pregnancy	Tests a urine sample for hormones that indicate a pregnancy, as soon as one day after a menstrual period is delayed
Ovulation	Predicts ovulation time by testing for hormones in urine of women who are having trouble conceiving babies
Blood Pressure	Allows people who are being treated for hypertension to keep a daily, continuous record of blood pressure readings so that treatment can be adjusted
Blood Glucose	Gives diabetics frequent measurements of sugar level in blood, so that drug treatment is more precise
Colorectal Cancer	Measures traces of blood in the stool that may indicate hidden tumors

PRESYMPTOMATIC DIAGNOSIS

That's a mouthful of a subtitle, but the term *presymptomatic diagnosis* helps redefine the new health care game. In the past, infectious diseases were the leading cause of death, but vaccines and antibiotics have virtually conquered these diseases. Today's major killers—heart attack, cancer, and certain defects of the newborn—generally strike with little or no warning. By the time the threat is diagnosed, effective treatment is often impossible. Traditional health care systems are geared to treating a threat after symptoms appear, not before. It is now recognized, however, that a new approach is in order.

New technologies are increasingly able to spot disorders or potential problems before they are life threatening; this is the field called *presymptomatic medicine.* In addition to the simple, inexpensive devices or kits for diagnosing and monitoring symptoms at home, recent developments in this field include

- highly accurate new lab tests based on the body's natural immunological system. The tests pinpoint specific disease markers that were undetectable a few years ago. It may soon be possible to quickly detect any disease by examining a "fingerprint" of the human body's thousands of proteins.

- complex clinical diagnostic tests that have become increasingly automated. New multisample instruments offer fast, low-cost body fluid analyses, often in a matter of a few minutes.

- genetic research (as described in Chapter 2), which gives physicians early clues to severe fetal disorders and offers parents the alternative of terminating pregnancies.

As mentioned earlier, death by certain diseases was practically wiped out during the period 1960 to 1980. For example, the mortality rate for kidney infection fell by 70 percent, tuberculosis by 85 percent, and influenza by 93 percent. But as our longevity increases, so do the cumulative health effects of diet, smoking, stress, lack of exercise, and other factors. The years that modern medicine has added to our lives provide more time for illnesses related to life-styles to develop. The good news is that these diseases can often be detected in their early, presymptomatic stages.

LIFE-STYLE ASSESSMENT PROGRAMS

As the name implies, life-style assessment programs permit the early evaluation of patients who are especially prone to a particular disease— the business-stressed executive with a family history of coronary disease, the heavy smoker who runs a high risk of lung cancer, and the overweight person in a sedentary occupation. Once identified, the risks involved can be reduced by changing people's life-styles.

Self-assessment by means of a computer offers an effective means to accomplish life-style analysis. Through a branching algorithm (a set of computer instructions that continually narrows the applicable choices to accomplish a particular task), the computer takes the individual through a lengthy "dialog" about health habits. Then the computer prints out a prediction of life expectancy, based on life-style analysis and other data. A sample dialog is shown in Figure 7-1.

The program also predicts what effect specific changes in life-style will have in terms of life expectancy. When an individual sees this information on a printout, it can have a dramatic effect.

THE NEWEST DIAGNOSTIC TECHNIQUES AND INSTRUMENTS

Once a medically treatable or preventable risk has been identified for an individual, the next step is using techniques and instruments arising from diagnostic technologies, a few of which are described here.

Fluorescence Bronchoscope

Lung cancer is one of the most preventable of all malignancies, but it is also one of the most difficult to detect and treat. By the time symptoms appear or the problem is detectable by conventional chest X rays, it is often beyond treatment. The sad result is that lung cancer cure rates are tragically low—only about 15 percent.

The recently developed fluorescence bronchoscope is expected to boost that figure significantly by revealing lung tumors while they are still small and operable, six months to a year earlier than can be seen now.

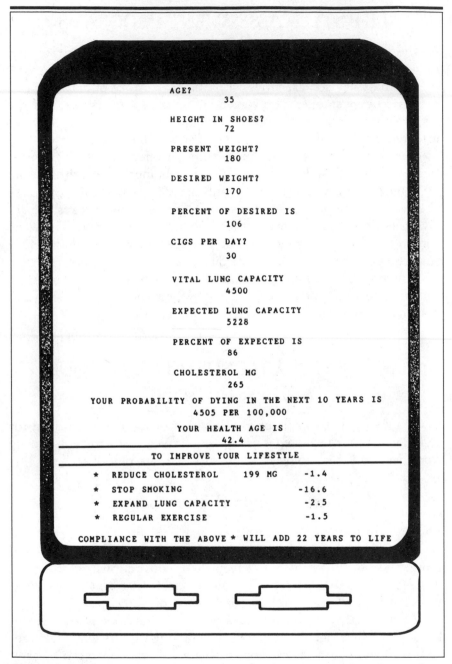

Figure 7–1 *Computer-Based Prediction of Life Expectancy Based on Life-Style Analysis*

Before using the bronchoscope, a technician injects the patient with a type of dye that is absorbed by malignant cells over a three-day period. The bronchoscope is then inserted into the lung and the tissues are illuminated with ultraviolet light from a krypton-ion laser. (I describe laser technology later in this chapter, but for now it is sufficient to understand that a laser is a device that focuses and intensifies light.) The dyed cancer cells fluoresce (give off light) under the ultraviolet illumination, and the light is transmitted to an image intensifier and video display. In this manner, malignancies only 2 or 3 millimeters wide can be detected (a millimeter is a thousandth of a meter or 0.039 inch). Because the process is invasive, that is, it involves the insertion of the bronchoscope and the injection of dye, this test won't be part of a standard screening, but rather it is used when physicians suspect a possible lung problem.

Mammograph

Breast cancer now kills more than 30,000 women every year. For women with a family history of this disease, early detection is vital. Fortunately, early detection is possible. Frequent periodic self-examination is the most efficient method for early detection, but according to the American Cancer Society, the number of women who practice this procedure is probably small.

Another simple and available method for early detection is mammography—low-dosage X rays of the breasts that detect tumors too small to be felt by hand. This screening procedure is recommended once every two years by the American Cancer Society for all high-risk women between the ages of 35 and 50, and for all women over 50.

Thermograph

Another device for the early detection of breast cancer is called a *thermograph* (or heat detector). This device detects the greater infrared radiation of tumors compared to normal tissue. The device consists of two fiber pads inlaid with rows of heat-sensitive dots. The pads are applied to the breast for 15 minutes, then examined. Skin temperature differences of only 0.5 degree Fahrenheit (0.27 degree Centigrade) cause the dots to change color. The resultant color pattern may alert the technician to the need for further examination.

Amniocentesis

An *amniocentesis* procedure can detect birth defects. It involves the withdrawal and analysis of a small amount of the watery fluid in which the human embryo is suspended in the womb. This procedure is usually performed during the 16th or 17th week of pregnancy and provides a good deal of information about the fetus. Fluid is withdrawn by needle after an ultrasound examination (described later) has located the position of the fetus in the womb. The fluid is then analyzed for a number of proteins and enzymes. Chromosomal analysis can reveal any genetic abnormalities.

Controversy is inherent in this technology. Amniocentesis and chromosomal analysis are now 99 percent effective in predicting birth defects. For this reason, many parents opt to end the pregnancy and avoid the emotional and financial hardships involved in raising handicapped children.

Critics point out that if a fetus can be aborted because of some devastating neurological defect or physical malformation, why will this technique not lead to abortion because of some relatively minor disorder, or even because of gender? Amniocentesis is an example of how the current technology is generating ethical questions.

IMAGING TECHNOLOGY

CT or CAT Scanners

Conventional X-ray radiographs, which view the body from only one angle, can be difficult to interpret because the shadows of bones, muscles, and organs are superimposed on one another. Also, large molecules such as calcium absorb X rays as they pass through the body, partially masking whatever lies behind them.

Computerized axial tomography (CAT), also called computerized tomography (CT), machines represent a major improvement. Using a thin, fan-shaped X-ray beam, a CAT scanner produces a cross-sectional view of tissues within the human body. Figure 7–2 illustrates this technique. CAT scanners view a "slice" of the human body from many angles by revolving an X-ray tube around the patient. The X-ray source is rotated rapidly—almost once a second in today's machines—making

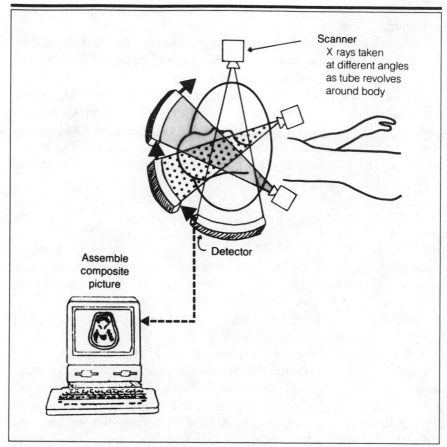

Figure 7–2 *Computerized Axial Tomography (CAT) Scanner*

hundreds of individual pictures. Detectors on the opposite side from the X-ray tube record what the scanner sees and transmit this information to a computer, which reassembles the data into thin cross-sectional slices.

CAT scanners can assess the composition of internal structures, discriminating between fat, fluid, and gas. A CAT scanner can also show the shape and size of various organs and lesions, and is sensitive enough to detect abnormal lesions as small as 0.039–0.078 inch (1 or 2 millimeters).

The scanners convert X-ray pictures into digital computer code that makes high-resolution video images. The computer graphics used are similar to those used to reassemble pictures beamed back from distant space probes.

CAT scanners are employed not only for detecting and diagnosing dozens of conditions but also for observing the effects of therapy and following up on the results of surgery.

Three-dimensional CAT images have recently been developed that can simulate the skin, muscle, and bone surface that physicians encounter during surgery. These images have also been used to evaluate brain tumors, head and neck malignancies, spine injuries, and pelvic fractures. The 3-D program is reported to reduce operative time and improve surgical results.

Sonography or Ultrasound

Seeing within the body with high-frequency sound waves is a painless, relatively inexpensive technique that has a wide range of uses. Sonography uses sound waves to look within the body. The system uses a piezoelectric (or pressure-sensitive) crystal that converts electric pulses into vibrations that penetrate the body, strike the organs within, and reflect back to the surface, where the crystal functions as a receiver. The time delays of these returning signals sketch the target's location, size, shape, and even texture for display on a screen. Sonography can obviate the need for CAT scanning and spares the patient from exposure to the mild radiation of other diagnostic techniques. Sonography is the only body-scanning technique recommended for pregnant women and is also useful in the examination of the breasts, heart, liver, and gallbladder.

Digital Subtraction Angiography (DSA)

Digital subtraction angiography is an imaging technique that produces clear views of flowing blood or its blockage in narrowed vessels. The term *digital* means numerical, and *angiography* refers to blood-vessel imaging, but what about *subtraction?*

This is how the subtraction works: First, a fluoroscopic image is made of the body area under study, enhanced electronically. Then that image is stored in the memory of a computer. The information is digitized—that is, the image is divided into a grid of *pixels* (or picture elements). Each of the several hundred thousand pixels is recorded with its relative brightness ranging from black to white. Once this preliminary "mask" image is recorded, a contrasting agent or dye containing iodine that is opaque to X rays is injected into the patient's vein. The shadow that this opacity creates enables doctors to see the flow of blood. A second, contrasting image is now made, highlighting the flowing blood as revealed by the dye. The computer then subtracts image one from image two, leaving a sharp picture of the blood vessels. The subtraction process involves comparing the two images, pixel by pixel.

DSA provides a new approach to revealing narrowing or other problems in blood vessels. Instead of requiring insertion of a long catheter (tube) into an artery as in conventional angiography, DSA permits a shorter catheter to be inserted into a vein. This procedure is safer for the patient, because the catheter is placed in a low-pressure venous system rather than the high-pressure arterial system. In addition, DSA requires only one-third the amount of dye necessary in the conventional angiogram.

With DSA, physicians now can see blood vessels as small as 0.039 inch (1 millimeter) in diameter without injection directly into the artery. In addition to helping detect constrictions, DSA also measures the rate at which blood diffuses into the heart muscles, giving the physicians a good indication of whether a heart attack is likely to occur.

One of the most important uses of DSA is in a therapeutic treatment called *angioplasty* (described later in the chapter under Balloon Therapy). DSA and angioplasty can preclude the need for some heart bypass surgery.

Positron Emission Tomography (PET)

PET is a major new addition to the doctor's imaging devices and it differs rather significantly from the CAT (computerized axial tomography) described earlier. CAT shows an organ's shape and structure—but not how it is functioning. PET provides metabolic portraits that reveal the rate at which abnormal and healthy tissues consume biochemicals. PET scans offer functional perspectives of biochemistry occurring within living tissue.

The word *tomography* comes from *tomas*, meaning a cutting or section, and *graph*, meaning write: tomography then is the taking of sectional radiographs in which the image of the selected plane remains clear while the images of all other planes are blurred or obliterated.

PET is a type of imaging that involves tracing the action of radioactive substances inserted into the human body. *Radiation* is energy given off by an atom. Most atoms are stable and nonradioactive. Those that are unstable or radioactive emit radiation. The radioactive substance used in PET scanners must be produced in a cyclotron located close to the PET scanner. A *cyclotron* is an accelerator in which particles are propelled in spiral paths by the use of magnetic fields.

This low-energy cyclotron produces isotopes of natural body substances such as sugar or glucose. *Isotopes* are radioactive elements. These isotopes are produced by accelerating charged particles in circles and then smashing them into a target. The radioactive isotopes have a short half-life, meaning that they lose their radioactivity within minutes of being produced.

The patient inhales or is injected with a small amount of the radioactive material and is then positioned in the scanning device. The device is shaped like a large donut (see Figure 7-3). The radioactive material emits positrons wherever it flows. *Positrons* are elementary particles having the same mass as an electron but having a positive charge. The positrons collide with electrons, and the two destroy each other, releasing a burst of energy in the form of gamma rays. These gamma rays the body emits are detected by a ring of crystals around the area being tested, causing the crystals to light up. A computer records the location of each flash and plots the source of radiation, translating that data into an image.

Using PET scanners, physicians and scientists can now look directly into the brain or heart—without surgery—and observe the biochemical reactions taking place. PET allows the observation of changes in brain chemistry after a stroke or the biochemical abnormalities in disorders such as Alzheimer's disease, Parkinson's disease, or schizophrenia.

PET also permits the observation of damage to the heart following a heart attack. The scanner reveals the precise region affected and gives critical information in revitalizing traumatized tissue.

Positron emission tomography is in its development stage and is currently being used as a research tool rather than a diagnostic technique. Its use for patients may soon be commonplace, however, as its potential is extraordinary.

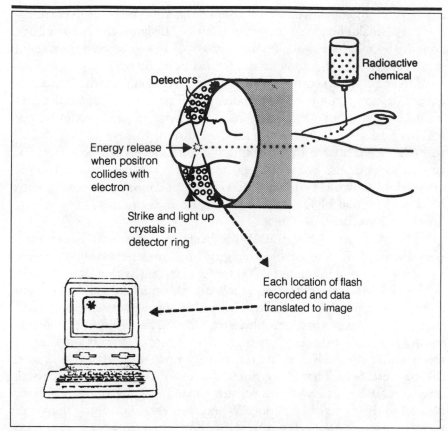

Figure 7–3 *Positron Emission Tomography (PET) Scanner*

Magnetic Resonance Imaging (MRI)

Magnetic resonance imaging, also called *nuclear magnetic resonance (NMR)*, may prove to be as great an advancement in modern medicine as x-ray technology was when it was invented in 1895. There are some 400 MRI machines operating in the United States today. Considering the cost of each machine (about $2 million), this dramatic growth since the technological development in 1977 indicates MRI's tremendous diagnostic ability.

MRI works on the principle that hydrogen atoms subjected to a magnetic field line up like soldiers on parade. When a radio frequency is aimed at these atoms, the alignment of their nuclei changes. When the radio frequency is turned off, the nuclei realign themselves, transmitting a small electric signal when they do. Because the human body contains a lot of hydrogen atoms, an image can be generated from these small electric signals, showing tissue and bone marrow.

MRI equipment consists of a huge electromagnet, a radio frequency generator, and a computer for evaluation. This setup is expensive and must be in a room completely insulated from external radio waves. MRI is considered superior to CAT scanning in several ways. CAT scanners use radiation that, although at a low level, can be a risk. For certain parts of the body, such as brain areas covered by thick bone material, MRI works better than CAT. No injections are necessary with MRI, and even at high magnetic power, no measurable adverse effects on tissue have been detected.

The MRI scanner surrounds the body with powerful electromagnets (Figure 7–4). Supercooled by liquid helium, these magnets create a magnetic field as much as 60,000 times as strong as that of the Earth. When the MRI magnets are turned on, the nuclei of the hydrogen atoms in the patient's body line themselves up with the field. This is similar to the way that iron filings can be made to line up pointing toward a nearby magnet. When radio waves of a certain wavelength are aimed at the body, the hydrogen nuclei realign themselves against the magnetic field. That is, the nuclei flip around to face in the opposite direction. When the radio waves are turned off, the nuclei flip back around to their original position. When they do this flipping they emit radio waves that are detected by the MRI receivers and are then used to create a computer-generated image.

MRI has proven to be a sensitive diagnostic tool for cancer. It is able to detect some brain tumors so small that they do not show up on X rays. The technique has also proven useful in measuring blood-flow rates from specific locations in the brain in order to identify stroke-prone patients. The effect of drugs on the brain can be determined by the use of MRI. Although MRI is expensive, the technology may permit the diagnoses of diseases earlier than would otherwise be possible, in which case, benefits will outweigh costs.

Of all the body-imaging diagnostic tools discussed in this chapter, MRI is considered to be the most promising, because it seems to combine the relative harmlessness of ultrasonics with the high-resolution imagery of CAT and PET scans. All of the imaging techniques discussed do help provide early diagnoses, and early diagnosis is most

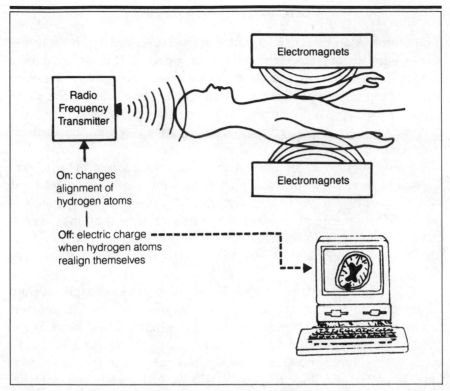

Figure 7–4 *Magnetic Resonance Imaging (MRI)*

important in the area of heart disease. This disease is still by far the nation's leading killer. Take a closer look now at this major threat to Americans' health and longevity.

THE HEART OF THE MATTER

About 1.5 million Americans will suffer heart attacks this year, and 500,000 will die as a result. Of these, approximately half the victims will not survive long enough to reach medical care. Don't get scared— just know your enemy. Before you learn about the application of high-tech medicine to heart problems, review briefly just how this life-sustaining pump works.

Functions of the Heart

The adult human heart is a muscular, four-chambered organ about the size of a grapefruit and weighing less than a pound. It is located almost exactly in the center of the chest. Its purpose is to provide the body with oxygen-carrying blood and at the same time eliminate the waste products for disposal. These tasks are carried out by two main pumping chambers—the right and left ventricles, separated by a muscular dividing membrane—and a series of one-way valves.

Oxygen-depleted blood from throughout the body enters the heart via two large veins. (Remember: Veins lead into and arteries lead out of the heart.) With each contraction of the heart muscle, blood is pumped from the right chamber through a valve to the lungs via the pulmonary arteries. In the lungs, the blood exchanges its carbon dioxide for fresh oxygen and returns to the heart through the pulmonary veins.

The heart's constant activity—about 70 contractions per minute during our entire lives—requires a steady oxygen supply. This demand is met by a series of arteries encircling the heart. These arteries are called *coronary arteries,* and they carry freshly oxygenated blood directly to the heart muscle. These arteries can become choked with fatty deposits and blood clots, gradually narrowing the internal diameter of the artery until the blood flow is limited or stopped altogether. The result is a heart attack and often sudden death. The heart attack is the most common dysfunction of the heart, but not the only one.

Fibrillation, which is the result of a defect in the heart's natural pacemaker, is another major source of trouble. When the heart fibrillates, it follows a jerky, irregular rhythm that fails to move blood through the heart in a normal manner. If regular contractions are not restored in a few minutes, death invariably results.

Now this is all pretty scary stuff, but high-tech medicine has made great progress in the last few years in fighting this threat. You already know about the new imaging techniques that can lead to early diagnosis and early treatment. New tools for *treating* heart disease also are developing at a rapid pace. Most of us know about heart bypass operations because they have been around for some time, but some of the newer techniques—electrical defibrillators, programmable pacemakers, synthetic replacements for the heart or its parts, balloon angioplasty, hemopumps—were not available until just a few years ago. These are some of the new technologies to know about.

Bypass Surgery

In this procedure, a portion of a superficial vein is removed from the patient's leg and cut into segments. One end of the vein segment is inserted into a small hole cut into the aorta (the body's main artery). The other end of the vein segment is inserted into the coronary artery downstream of a life-threatening blockage. This creates a new route for the blood to reach the heart muscles, bypassing the trouble spot. Most bypass operations involve two or more of these grafts (hence the terms *double* and *triple bypass* operations). Bypass operations do not eliminate the underlying disease, however, and a significant percentage of patients develop new blockages in a short time.

Variable-Rate Pacemakers

The human heart is equipped with its own pacemaker, called a *sinus node*, which produces electrical impulses that cause the heart muscle to contract and pump blood throughout the body. When the natural pacemaker fails for any number of reasons, fatigue, dizziness, fainting spells, or other symptoms may result. If medications fail to restore the function of the natural pacemaker, an artificial electronic pacemaker can be used to restore the proper heartbeat.

Over 110,000 pacemakers are installed every year in patients in the United States. Most recipients are victims of something called *bradycardia*, which is an abnormally slow heartbeat (less than 50 beats per minute) that fails to meet the body's energy needs. A heartbeat that is too fast (a condition called *tachycardia*) is a more common problem and is treated largely with medication.

The modern pacemakers installed today differ in important ways from earlier pacemakers. First, they are much smaller, weighing less than two ounces, and second, the installation procedure is now relatively simple. The newer power source, usually a sealed lithium iodide battery, can supply energy for up to a decade. Whereas early pacemakers were preset to maintain the heartbeat at a fixed rate—usually between 60 and 80 beats per minute—today's units can track the body's demand for oxygen and either step up or slow down the pace. Finally, modern pacemakers may be adjusted and monitored externally. The physician uses radio-frequency signals to fine-tune a patient's unit to meet individual needs. In some cases these signals can be transmitted over a telephone line, eliminating the need for visits to physicians' offices for pacemaker adjustments.

Automated Defibrillators

New devices that automatically detect and interpret abnormal heart-beats and deliver electrical jolts to restore the normal rhythm could save many Americans who die each year of cardiac arrest. The machines are advanced versions of the defibrillators that doctors and paramedics have used for years. The devices are so automated that only a modest amount of training is required to use them. They could be especially useful in rural areas or other places where trained paramedics may not be available.

When attached to a victim's chest, the device automatically checks the victim's heart and delivers a shock if it finds that the victim has suffered a cardiac arrest. If given soon enough, the shock almost always restarts the stalled heart.

How soon is soon enough? In cardiac arrest, the heart stops pumping blood because of a disruption in its electrical signals, and the victim falls unconscious. Frequently, the first to respond to an emergency call are firefighters or the police, who administer cardiopulmonary resuscitation (CPR) but not electrical shocks. When the heart has stopped, the best CPR in the world cannot substitute for the needed electrical shock. Unless the heart starts beating again within four to eight minutes, brain damage or death is likely.

The new automatic defibrillator has been tested by Seattle fire-fighters. After only four hours of training with the new device, fire-fighters were able to give shock treatment to victims who otherwise would have had to wait for the arrival of a hospital paramedic team. The result has been a significant increase in the survival rate of cardiac arrest victims.

Synthetic Spare Parts

In many heart patients the disease is limited to the left chamber of the heart—the powerful left ventricle that forces oxygenated blood into the aorta and out to the extremities. A technical answer to this problem is called a *left ventricular assist device (LVAD)*. This is a small air-driven device that can be mounted externally, near the heart. The device takes over the work of the left ventricle for up to two weeks. The left ventricle still contracts during this time, but its load is lighter because the blood is routed through the assist device. This device provides time for the physicians to take whatever action is necessary for long-term therapy.

New synthetic heart valves are now being used to replace worn-out or defective human valves. Natural valves are often damaged by rheumatic fever. This infection often leaves behind scar tissue that prevents valves from opening and closing properly.

There are two basic types of artificial valves: the totally synthetic models and those fashioned from animal heart tissue. Each type has advantages and disadvantages. Mechanical valves are durable but noisy—the "click" sound may be distracting not only to the patient but to persons nearby. Animal valves are prone to calcification and failure. The latest in synthetic valves uses leaflets made out of a poly-etherurethane that is structurally and biologically similar to natural body tissues.

Balloon Therapy

Balloon therapy—more technically known as *percutaneous balloon angioplasty* (*percutaneous* meaning through the skin, and *angioplasty* meaning blood-vessel repair) is today becoming a major alternative to heart surgery. In this procedure, a catheter with a tiny plastic balloon is threaded into the artery and, guided by a fluoroscope, maneuvered through the body until it reaches the blockage. There the balloon is inflated, slowly compressing the fatty material (called *plaque*) against the artery walls and thus increasing the diameter of the blood vessel.

With improved and smaller catheters, angioplasty is being used to open clogs in arteries of the arms, legs, and other areas—in some cases as an alternative to coronary bypass surgery. It has been adapted recently to opening up diseased heart valves in children—eliminating the need for open heart surgery.

Hemopump

This device is a tiny artificial heart pump about the size of an eraser on a pencil. The hemopump is inserted through an artery in the patient's leg and then pushed up into the heart itself. The pump is powered by a small electric motor that sits next to the patient's bed. Minute turbine blades in this miniature pump pull blood out of the left ventricle through a tube into the aorta. The hemopump can sustain a failing heart for several days while physicians work to stabilize a patient's condition.

As a result of the technologies just described, as well as other new technical developments, and most importantly, recognition of the importance of life-style changes, heart disease may in the near future forfeit its title as America's number-one killer.

MEDICAL LASERS

Lasers, which are devices that focus and intensify light, are widely used in medicine today. Among the many uses for surgical lasers, for example, are the spot-welding of detached retinas in the eye, vaporization of abnormal growths and tumors, and halting of internal bleeding. New developments in laser technology are opening up even wider areas of use, and laser therapy may become the standard procedure for the future. An understanding of the fundamental principles will help you evaluate lasers' potential.

Laser Technology

Laser is an acronym for light amplification by stimulated emission of radiation. Basically, a laser is a way of making pure or "coherent" light. Ordinary light from a light bulb or from the Sun consists of many different colors (wavelengths) going in all directions. The light is thus incoherent and spreads and scatters over distance. If you aim a flashlight into the night sky, the beam will not travel far before it disappears. But a pencil-thin laser beam can stay strong enough to travel the nearly quarter of a million miles of space between the Earth and the Moon.

A laser compresses light into a thin beam of one wavelength. A laser machine consists of a glass rod or tube filled with a gas. When the laser is "pumped" with energy in a variety of ways, electrons in the gas are excited into higher energy states. The high energy electrons release their extra energy as light. That light is amplified as it bounces back and forth between two mirrors. The laser light that is emitted from the machine is a single color (which depends on the gas used) and is a narrow, concentrated, powerful beam.

Conventional lasers use various substances such as carbon dioxide or even gold to produce differing wavelengths of laser light. A new type of laser, the *free-electron laser*, uses a magnetically controlled electron beam, similar to the beam in a television picture tube, to produce a range of wavelengths. Free-electron lasers are 20 times more efficient than previous laser machines, and they are more powerful than conventional laser beams.

A host of medical applications are foreseen for free-electron lasers in the next 5 to 10 years. These include scalpel-less procedures to destroy deeply embedded tumors, improved bone surgery, and the killing of AIDS or other viruses.

Laser Therapy

An argon gas laser can be used to produce a wavelength that is absorbed by red pigment. It can pass harmlessly through the eye's transparent lens and the clear fluid of the inner eye. But when the beam hits the retina (that part of the back of the human eyeball that receives the images produced by the lens), the laser causes coagulation of the blood vessels, fusing adjacent tissues. By this means, eye surgeons can "spot weld" detached retinas.

Doctors are currently testing a dramatic new laser surgery that they hope will correct vision problems and eliminate the need for eyeglasses in many people. This procedure employs something called an *excimer laser* to sculpt the cornea—the transparent front outside coating of the eye—correcting nearsightedness, farsightedness, and astigmatism.

Excimer lasers generate pulsed ultraviolet beams that do not heat tissue, as other lasers do. Instead, the beam breaks chemical bonds and allows cells to be washed away without damaging surrounding tissue. Doctors hope that this approach will let them reshape the curve of the cornea without scarring. The shape of the cornea determines how light is focused on the retina in the eye.

Laser Angioplasty

The previously described balloon angioplasty works best when a blood vessel is only partially blocked. Total blockage usually requires coronary bypass surgery. A new procedure incorporates both the balloon and a tiny laser and is designed to be used on totally blocked vessels. In this procedure, the laser beam is intended to vaporize the plaque, cutting a channel. Then the deflated balloon is threaded through the channel and inflated to push the remaining plaque against the vessel walls. Laser angioplasty is still in its experimental phase, but it has been approved for blood-vessel blockages in the legs.

BIONICS

How close are researchers to developing the bionic man of TV fame? Artificial organ development—or *bionics*—is continuing at a rapid pace. The National Institutes of Health estimates that several million artificial implants are now being used each year in the United States alone. Although the results of mechanical heart transplants were not promising, other remarkable spare parts have been emerging: artificial ears, synthetic skin, artificial blood vessels, a pocket-sized artificial kidney, and artificial joints and limbs. Bionic development has almost turned television science fiction into medical reality.

Somewhere, researchers are working on an artificial substitute for almost every organ in the human body. In addition to mechanical/electrical parts, scientists are trying to create hybrid organs that combine living tissue with artificial parts. Currently in development are hybrid kidneys, livers, and pancreases.

Artificial joints (hip, shoulder, elbow, knee, and ankle) can now be made to order. Bioengineers use a reference bank of implant designs, special software, and a computer terminal to design joints to meet a specific patient's needs. They fit the design on the screen to X rays and cross-sections of the patients' bones. After the design is completed, the computer then feeds the data into various machines that cut the implant from materials such as stainless steel, polyethylene, and chromium. At the present time, the costs are high. A new knee, for instance, costs about $5000.

New technology has also created a new generation of artificial limbs for amputees. For example, an experimental artificial leg uses an electromechanical knee controlled by the will of the wearer. When the wearer sends signals to the muscles above the amputation site, a computer connected to electrodes on the thigh senses the muscle activity and sets the leg in motion in what appears to be a natural movement.

A variation of this technique is being developed for artificial elbows, arms, and hands. Electrodes pick up signals from the brain to the remaining arm muscles, enabling the elbow or arm to respond naturally. An artificial hand can be controlled by the movements of muscles in the opposite shoulder by means of a small cable and switch.

FIBER OPTICS

The use of flexible fiber-optic instruments to view internal organs of the body has in recent years changed physicians' approach to gastrointestinal disease. Using hair-thin glass fibers to provide light and allow visualization, it is now possible to directly inspect the esophagus, stomach, duodenum, colon, and abdominal cavity without surgery. Called *endoscopy*—viewing internal organs of the body with an instrument—this technology was originally concerned mainly with diagnosis, but in recent years it has been increasingly used for treatment as well. One current therapeutic use of fiber optics is the removal of colon polyps, tumorous growths that are generally considered precursors of colon cancer.

The extraction of foreign objects from the stomach and esophagus—fish bones, toothpicks, pins, and buttons, for example—can now be done using fiber-optic endoscopy. Devices that can grasp a foreign object and remove it are now available thanks to fiber-optic developments.

MICROSURGERY

Microsurgical techniques developed along with improvements in microscopes and illumination technology. Today surgeons can use binocular operating microscopes powerful enough to provide a three-dimensional view of the field of operation at magnifications ranging up to 40 times. Also available are microscissors, miniature probes, hooks, clips, and suture needles.

Ophthalmologists were among the first to make use of microsurgical techniques in performing delicate eye operations, including transplants of the cornea (the window of the eye). Ear surgeons have also made good use of microsurgical techniques. An example is the reconstruction or repair of the tiny bones in the middle ear that transmit sound vibrations to the inner ear.

Microsurgical bone transplants are now possible that would not have been attempted a few years ago. Connecting an implanted bone section to the surrounding blood vessels requires microsurgery, but this technique greatly increases the chances of success in any kind of implantation surgery.

TRANSPLANTS

Transplantation of Brain Tissue

Transplantation of an entire human brain, a favorite subject in science fiction, may never be possible in reality, but transplantation of a portion of brain tissue is a promising field for the treatment of Parkinson's disease and other neurodegenerative (nervous system deterioration) problems.

Parkinson's disease is a major brain/nervous-system disorder that affects more than 1 million Americans and that ultimately results in severe disability in the majority of cases. Whereas treatment for most patients at an early stage of the disease (using the drug levodopa) is good, treatment for those at a late stage of disease has been a problem.

Because most of the symptoms of Parkinson's result from the loss of certain neurons in the brain, and because of the limitations of the drug levodopa as a substitute for that loss, researchers have begun to consider replacing the lost tissue with transplants.

Transplantation of Human Fetal Tissue

Parkinson's involves the slow degeneration of brain cells that secrete *dopamine*, a movement-coordinating chemical. By transplanting fetal cells able to produce dopamine, medical researchers have succeeded in relieving the symptoms of Parkinson's. The use of human fetal tissue, however, raises ethical issues.

Research involving human fetal tissue has been the subject of intense political debate in this country for 20 years, and the recent use of fetal tissue in transplantation continues the controversy. The Reagan and Bush administrations' view of the moral status of the human embryo limited federally funded research on fetuses. Current federal regulations, however, do permit the use of tissue from dead fetuses in experimental transplantation when conducted in accordance with state law. Nonetheless, ethical concerns remain, and there has been continued political resistance to funding fetal tissue research with public money.

The controversy centers on the abortion issue. The primary argument of those who object to federal funding of fetal research is that such funding would "institutionalize" government complicity with the "abortion industry." There has been little objection to the use of spontaneously aborted (that is, miscarried) fetuses. The greatest outcry seems to have been against the use of electively aborted fetuses.

Why not limit transplantation research to fetal tissue from spontaneous abortions and avoid the ethical problems involved in induced abortions? Unfortunately, the issue cannot be sidestepped that way. It is now well established that about half of the fetuses spontaneously aborted in the first trimester and 20 percent of those aborted in the second trimester are chromosomally abnormal. In addition, a variety of harmful microorganisms have been associated with spontaneous abortions. For these reasons, tissue from spontaneously aborted fetuses is not recommended for transplantation to human subjects. Fetal tissue transplantation will continue to be a controversial issue and for this reason, alternatives are being explored.

Using a Patient's Own Tissue for Transplants

The possibility of transplanting dopamine-secreting cells taken from a patient's own adrenal gland atop the kidney has been explored. Using a patient's own tissue avoids problems of host graft rejection and the necessity for using drugs to suppress rejection. *Graft rejection* occurs when there is an immunological reaction against a transplanted organ or tissue because of a poor genetic match. Using a patient's own tissue also bypasses ethical issues involved in transplantation of fetal tissue.

Unfortunately, these *adult adrenal autografts* do not reverse parkinsonian symptoms as well as do fetal cell grafts. In fact, adrenal tissue transplants have had overall disappointing results.

Use of Animal Tissue

Other possibilities being explored at present involve the use of animal fetal tissue. It is possible that monkey fetal cell tissue could be altered for use in humans. Donor brain tissue might also come from baboons or chimpanzees. Interspecies grafts have worked between mice and rats. Brain-cell implants in rats with symptoms similar to those of Parkinson's disease have succeeded in relieving the symptoms.

Synthesized Tissue

Fetal tissue is thought to have unique properties, and if researchers can identify what gives these cells their regenerative ability, ways may be found some day to synthesize the substances or produce them in laboratory cultures. All of the grafting techniques discussed here are considered crude, but neurologists are convinced that, in the near future, procedures will be developed using purified, enriched, or genetically engineered cells that will produce the specific results desired.

Brain tissue transplants are allowing neurologists to play a more active role in treating brain/nervous system disorders. In the past, neurologists could describe and catalog disorders and offer solace and palliative treatment to patients. In the future, neurologists will have to be able to select the patients and disorders that may be treatable by transplantation.

DESIGNER DRUGS FOR THE BRAIN

Better understanding of the action of neurotransmitters, the brain's chemical messengers, is the future for neurology. About 50 different neurotransmitters have been identified so far, and scientists estimate that there are at least 200 in all. Present-day research centers on expanding the list of known transmitters, finding their precise location in the brain, and determining the mechanism of regulation.

Each transmitter "fits" into a special receptor on the surface of a brain cell. Understanding how these transmitter/receptor pairs work is leading the way to the development of specific drugs that can be applied directly to an intended receptor. By designing a drug for a specific receptor only, undesirable side effects can be avoided. New, nonaddictive pain killers are one example of possible results. Other designer drugs are experimental compounds that increase attention spans, improve memory, and prevent suicidal depressions.

The potential for designer drugs is far reaching. Drugs to curb aggressive tendencies, dispel fear, produce calm, or affect just about any other emotion are possible. If the brain is just a little box with emotions packed into it, as some neurologists say it is, and the bio-chemical correlates of each emotion can be controlled, the potential (for good or ill) is tremendous.

POSTSCRIPT

This chapter, of course, only touched on the high spots of state-of-the-art technical medicine. I hope the chapter piqued your interest and curiosity about a subject that can change your life. Medical science is dynamic. To understand what our doctors are recommending for us, the costs and the controversies, or even merely to be able to ask the right questions, we will have to keep abreast: to know and understand as much as we can.

KEY CONCEPTS

▶ Medical diagnostic technology has undergone a major breakthrough in recent years, leading to earlier detection and more effective treatment of illness. However, the years that modern medicine has added to our lives provide more time for illnesses related to our life styles to develop.

▶ Imaging technology such as CAT, sonography, DSA, PET, and MRI permit physicians to peer into the human body as never before.

▶ New technological tools and techniques for treating heart disease are proliferating at a rapid pace. Most of these—electrical defibrillators, programmable pacemakers, synthetic replacement parts, balloon angioplasty, or hemopumps—were not available until just a few years ago.

▶ Laser technologies—involving the focusing and intensifying of light—have great medical potential. Laser surgery and laser therapy may become standard procedures in the near future.

▶ Fetal tissue transplants to repair aging or damaged brain tissue offer the best hope at the moment for Parkinson's disease (and even Alzheimer's disease), but the procedure is plagued by ethical problems. In the longer term, synthesized tissues or designer drugs may offer the best potential for brain/nervous system malfunctions.

8

Getting There from Here:

TRANSPORTATION TECHNOLOGY

CARS, PLANES, AND TRAINS, that's how we get around these days. (Oh, we take a boat now and then but that's mostly for fun.) What about the technologies involved in these means of transportation? Do we really know what makes them go?

Take cars, for instance. How many drivers understand the basic engineering principles governing engine operation or the means for transmitting power to the wheels? How important is aerodynamic drag to overall performance of our cars? Should we be considering alternative fuels or even alternative engines? What is the realistic potential for electric or hybrid cars? How is future technology likely to transform our cars?

Next, take planes. When the British-French Concorde made its first flight more than a decade ago, this sharp-nosed supersonic jet that could fly across the Atlantic in under four hours was the last word in high-tech passenger planes. That is about to change. On the drawing board are the high-speed civil transport aircraft with a capability of flying at speeds close to Mach 3, or three times the speed of sound (at sea level, the speed of sound is 740 miles per hour). The Mach 5.5 "Orient Express" is designed to fly 300 passengers across the Pacific Ocean in two hours. Finally, the hypersonic NASP (for National Aero-Space Plane) is designed to fly into orbit. Should we plan our future vacations around one of these concepts? A review of the technologies will help us decide.

The term high-tech train has, for many of us, been an oxymoron. Train technology, at least in this country, was stopped in its tracks many decades ago, and progress is not the word often used in connection with passenger train travel. There are those, however, who think this state of affairs is about to change. Those favoring a renewal in train transport anticipate a passenger-rail renaissance in the United States. They point to the growing congestion on the nation's roads and at its airports, and to the pollution caused by petroleum-burning engines. Can the success the Japanese, French, and Germans have with high-speed trains be duplicated in this country? Should we give trains another chance?

ADVANCED AUTOMOBILE TECHNOLOGY

A friend of mine once announced that he spoke three languages: American, Baseball, and Cars. Car buffs understand this completely. They devour the latest car magazines, attend the annual car shows, and in general can't get enough of the subject. The rest of us have a tendency to fall behind, and we have to play a little catch-up to understand what's happening to our favorite form of transportation.

Fundamentals

Fundamental scientific laws govern the ability of an automobile engine to convert the energy in the fuel to a form that is able to propel the car down the road. These are referred to as the First and Second Laws of Thermodynamics. *Thermodynamics* is that branch of physics that deals with the transformation of heat into work and other forms of energy. The First Law simply states that "energy is conserved"; that is, it's indestructible—there is always the same total amount of energy in the universe. Energy is neither created nor destroyed, it just changes form, such as from chemical energy in fuel to heat or mechanical energy.

The Second Law is a bit more complex. The Second Law states that "the entropy of the universe tends to a maximum." *Entropy* is a measure of the total disorder, randomness, or chaos in a system. The effect of increased entropy, then, is that things progress from a state of relative order to one of disorder. With this progressive disorder there is increasing complexity.

Every time we convert energy from one form to another we lose on the deal. Some of the energy is wasted. It is not lost—that would be contrary to the First Law; but it is converted to heat that is dissipated in the environment. The portion of the energy that is unavoidably dissipated as nonuseful heat is reflected in the measurement of entropy. In our automobiles, this heat rejection occurs mostly through the cooling system's radiator and the exhaust pipe. In fact about 70 to 80 percent of the energy in gasoline flows out of the automobile in the form of rejected heat. Much of this is accounted for by the Second Law.

How does the Second Law affect our automobile engines? There are two types of engines in common use today, the Otto cycle, and the diesel. The "four-stroke" engine powering most of our cars today was invented by Nicholas Otto; hence the name. I discuss the differences between these engines further on, but the Second Law affects them all.

The specific effect of the law on the efficiency of an engine varies with the compression ratio and the ratio of air to fuel. The compression ratio is the ratio of the combustion chamber volume with the piston at the bottom of the chamber versus the piston at the top. A typical gasoline engine today has a compression ratio of 8:1 or 9:1 and an air-to-fuel ratio of 15:1. That means 15 pounds of air enters the engine for every pound of fuel. A typical diesel engine has a compression ratio of about 20:1 and an air-to-fuel ratio typically much leaner, about 18:1 (more air and less fuel) than gasoline engines.

Today's gasoline limits us to compression ratios around 10:1, and gasoline engines are difficult to keep running with air-to-fuel ratios more than about 20:1. The theoretical (Second Law) maximum efficiency—that is, the efficiency with which the chemical energy inherent in the fuel is transformed to the mechanical energy needed to propel the car—is 40 to 45 percent, and that is only achievable if all other losses (such as from friction) are eliminated. A more reasonable potential efficiency for gasoline engines is 35 percent, which is substantially better than today's engines, which achieve about a 25 percent efficiency in typical driving.

Today's automobiles have reciprocating internal combustion engines. That is, they have pistons that go up and down and combustion occurs intermittently. A combustible mixture enters the cylinder chamber, is ignited by the spark plug, burns to produce energy that drives the piston, and then leaves the chamber as exhaust before the process is repeated.

The Otto cycle is by far the most common engine in use today. This engine has four distinct strokes: intake, compression, power, and exhaust. Each stroke represents a sweep of the piston from top to bottom or vice versa, and each stroke thus represents one-fourth of the cycle. The intake downstroke sucks in a new charge of air and fuel. Then the upstroke compresses the charge. Near the top, the spark plug ignites the mixture, and, because it expands as it burns, the piston is forced down, producing power. The exhaust stroke purges the cylinder of the spent gases. (Figure 8–1 illustrates Otto cycle engine operation.)

Two-stroke engines offer some advantages, and major auto makers have been working to develop the technology. In the efficient two-stroke design, the piston moves up and down only once between firings. Today's two-stroke engines are smaller and run more smoothly than four-stroke engines of comparable power. Small two-stroke engines have been used for years to power machines from chain saws to outboard motors. Recent breakthroughs in fuel injection have created two-stroke models that are clean enough for use in cars. Two-stroke engines use less fuel, and their light weight helps to make front-wheel-drive cars better balanced and easier to steer.

An internal combustion engine is essentially an air pump, and significant improvements in efficiency result from streamlining the flow of fuel and air into and out of the cylinders. Turbocharging and multiple-valve technologies are two approaches to making engines breathe easier.

A turbocharger consists of a turbine in the exhaust stream that is connected to a compressor in the inlet stream. Turbos are air compressors, pressurizing the intake charge to force more air into the combustion chamber. The result is increased power, but as usual, there is a cost involved. Turbos are expensive to manufacture, and because they spin at over 100,000 rpm, special materials and high-precision manufacturing are required. The advantage of turbocharging is the ability to get the same amount of peak power and acceleration from a smaller engine. Turbochargers will continue to be used on high-performance cars where superior acceleration is important, because their

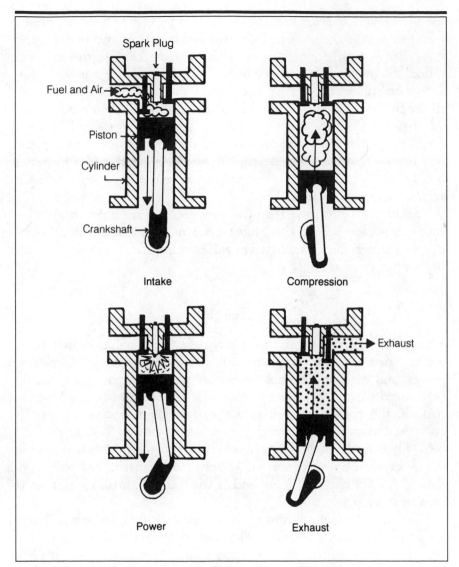

Figure 8–1 *Four-Stroke Otto Cycle Engine Operation*

increased complexity and cost are offset by the fuel economy obtained compared to that for a larger engine.

Multivalving increases power without the complexity of turbocharging. A conventional engine has two valves in each cylinder—one for intake and one for exhaust, as shown in Figure 8–1. A popular trend in engine design is to have two intake valves and two exhaust valves in each cylinder. Four-cylinder engines that use this configuration are known as 16-valve engines. This allows better breathing (just as two nostrils are better than one), which means there is less resistance to the fuel-air charge entering the engine. As a result, more power can be obtained from the engine. This is a big advantage at full throttle when the driver wants the most out of the engine. Just as in the case of turbocharging, a design feature that adds power results in better fuel economy because a smaller engine can be used to obtain the same acceleration performance.

Multivalves cost more than the conventional arrangement, but because they allow a lighter and more efficient package than simply putting in a bigger engine, multivalves will continue to appear on a greater number of engines.

Diesels

The diesel engine is similar to the typical Otto engine in that there are pistons that carry out the same four strokes: intake, compression, power, and exhaust. The differences are that the inlet charge has no fuel mixed with it and there is no spark plug to initiate combustion. Instead, the fuel is injected into the cylinder near the top of the compression stroke, and a high compression ratio is used so that the charge becomes hot enough to ignite spontaneously. This is similar to the detonation phenomenon (or knock) that is considered objectionable in the Otto cycle engine and is the basic limit on that engine's compression ratio.

The diesel's main advantage is its superior fuel economy. There are some drawbacks, however. The high pressures in diesels are hard on engine components and generate more noise than does an Otto engine. Diesels must be built heavier and sturdier than their gasoline cousins to allow them to withstand the beating. Diesels also have a significant exhaust problem, although considerable research is underway to reduce diesel soot and other undesirable emissions. At this writing there are few diesel passenger cars being sold in the United States, but should future fuel prices increase, the diesel engine production lines may be opened again.

Wankel Engines

Wankel engines are Otto cycle engines with a major difference. The pistons are replaced with a rotor. The cycle is the same. The main advantage to the Wankel is that it can produce more horsepower per pound of engine than a conventional engine. However, the rotor configuration limits the compression ratio and requires many sliding seals, which sometimes leak, thus causing more emissions and lower efficiency.

Displacement

Before you leave the engine basics, consider the terminology used in automobile ads and brochures. These ads are confusing, if not misleading. Most automobile brochures today refer to engine displacement (engine size) in liters, but some use cubic centimeters or even cubic inches. They are all referring to the same thing: displacement, or the volume of space through which the piston travels during a single stroke in an engine. The numbers you see in the literature refer to the space traveled by *all* the pistons in a single stroke. That is, a six- or eight-cylinder engine has a larger displacement number than a four-cylinder engine with the same size pistons.

Because a liter is equal to 1000 cubic centimeters (expressed as 1000 cc), that conversion is an easy one. For example, a 1.9 liter displacement is the same as a 1900 cc displacement. Manufacturers of subcompacts often advertise displacement figures in cubic inches (probably to make comparison difficult). Fool them. To convert displacement expressed in cubic inches to liters divide by 60. (Example: 173 cu. in. equals 2.8 liters).

Alternate Fuels

Because almost every city in the United States is now in violation of the Clean Air Act standards, Congress is considering new measures to reduce the air pollution caused by automobiles. One approach is to encourage the use of fuels other than gasoline. A bill introduced by Representative Henry Waxman requires those areas with the worst air pollution to establish extensive alternate fuel programs so that 30 percent of new cars in those areas would be operating on cleaner-burning ethanol, methanol, or natural gas by 1998.

Both Ford and General Motors plan to provide the state of California with test fleets of "flexible fuel" cars, able to run on methanol, ethanol, or gasoline, in the next few years. The flexible-fuel cars in the test fleet would emit 50 percent fewer smog-producing chemicals than comparable gas-only cars, according to the California Energy Commission, which is subsidizing the test fleet.

To provide fuel for the test fleet, both Chevron and Arco have agreed to install methanol pumps at 50 service stations throughout California by the end of 1989. None of the suggested alternative fuels is as efficient as gasoline nor are they cost competitive at current prices. It is only through the use of subsidies that alternative fuels can be made cost-effective at this time, but the growing problem of dirty air may make some sort of subsidy necessary eventually.

Aerodynamics

Because the industry has gone about as far as it can go in downsizing cars, improved aerodynamics will have to play the major role in reducing the resistance to motion in future passenger cars. The aerodynamic drag of a car at a given speed is determined by the frontal area and the drag coefficient. Typical frontal areas are between 17 and 20 square feet. With other factors being equal, the car with the smallest frontal area gets the best mileage. However, the need to seat passengers comfortably limits the ability to reduce the frontal area of cars. That brings us back to the coefficient of drag, usually expressed as C_D.

Air resists the movement of a vehicle passing through it. The resisting or "drag" force increases with the square of the vehicle speed: Twice the speed produces four times the force. The engine power required to overcome the drag force increases with the cube of the vehicle's speed: Twice the speed requires eight times higher power. As you can see, even small reductions in drag can result in significant decreases in power requirements at high speeds. Figure 8-2 illustrates the forces and power necessary to overcome air resistance.

The average car today has an approximate C_D of 0.4 and a range between 0.25 and 0.54. Researchers have concluded that a drag coefficient of 0.15 is about the lowest level achievable in a wheeled vehicle moving over a road. Figure 8-3 shows the C_D of typical cars over the past years as compared with today's average and tomorrow's potential.

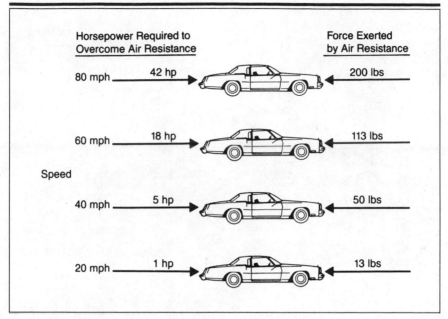

Figure 8–2 *Power Required to Overcome Air Resistance*

Electronics

Electronic components have already proven themselves in auto power-trains. Now they are emerging in braking, steering, and suspension systems. The result is a significant improvement in the safety and handling of U.S. cars and imports to this country. The dollar value of the electronic components in a 1985 car averaged $585. This figure is predicted to increase to about $1500 by 1995. Because the costs of electronic items drop with time, there could be 5 or 10 times more electronic components in tomorrow's car compared with today's. But what about the all-electric car?

Electric Cars

For decades the popular science magazines have been telling us that electric cars are just around the corner. It hasn't worked out that way, and the reason is technical—more specifically, battery technology. Not

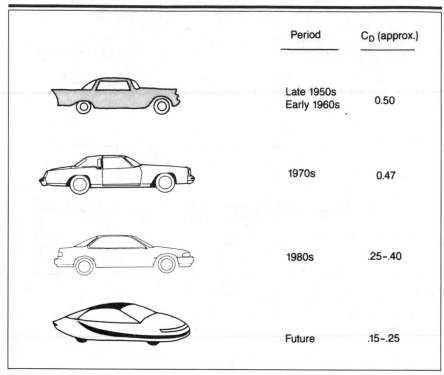

	Period	C_D (approx.)
	Late 1950s Early 1960s	0.50
	1970s	0.47
	1980s	.25-.40
	Future	.15-.25

Figure 8–3 *Drag Coefficient (C_D) of Typical Cars Over the Past Years Compared with Today's Average and Tomorrow's Potential*

that electrics don't have many advantages; it's just that at this writing the disadvantages far outweigh the virtues.

Electric cars provide an on-demand power system (no fuel-wasting idling) that is quiet and nonpolluting. With electric cars, there is freedom from the dependence on gasoline that was such a problem in 1973–1974. With electric vehicles, everybody's garage's electrical plug becomes the only "gas station" needed. The cost per mile of running an electric car is usually quoted as significantly less than a gasoline powered vehicle (but there is a hidden kicker in these calculations that you should take note of).

The main disadvantage of electric cars lies in their relatively terrible performance in speed and range, and in their "real" operating costs. Whereas any gasoline car can go at least 250 miles at 55 miles per hour (mph), the range of electrics for sale today is probably around 30 miles at 55 mph. I say *probably* because nobody ever quotes the range at 55 mph. Manufacturers' literature usually points out that an electric car has a 100-mile range and a maximum cruise speed of 55 mph, without mentioning that you can't do both. To achieve the 100-mile range you have to drive considerably slower than 55 mph.

The second major disadvantage of electric cars is the relatively inefficient state of battery technology. Today's batteries wear out after only a few years, and the cost of replacing the entire battery pack is about the same as putting a new engine in a gasoline car every few years.

In spite of the serious performance and cost disadvantages of today's electrics, there is room for optimism. There is real potential for big improvements in battery technology in the future, and a breakthrough in energy storage device technology could lead to a new advantage for electric car adoption.

General Motors' experimental solar-powered Sunraycer car was an interesting stunt but do not look for a solar-powered car in your local dealer's showroom for some time to come. Of more immediate interest is a solar-powered battery charger. Available now, the solar charger is a thin solar panel, about 7 inches by 13 inches, mounted in a plastic frame. You can place it on your dashboard or rear window deck. It will continuously charge your car's battery while your car is parked or driven in the sun.

Steering and Suspension

Among the most significant of recent innovations is four-wheel steering. This makes the car more maneuverable at low speeds and more stable during high-speed cornering. Other advancements are adjustable suspension (which automatically changes a car's spring characteristics to suit immediate driving needs) and all-wheel drive. All-wheel drive differs from four-wheel drive in that the former are full-time systems, permanently splitting drive torque between front and rear wheels.

All-wheel drive passenger cars use a center differential to divide power between front and rear. In conventional rear-drive cars, the differential is used to split torque between the two drive wheels, allowing the two wheels to turn at different speeds when the car is cornering. The inside wheel, traveling less distance than the outer one, can turn more slowly. In an all-wheel drive car, the differential distributes tractive force to the wheels with the best grip on the road, greatly reducing the chances of getting stuck because of ice, snow, or mud.

Safety

Two new developments in auto safety are worthy of attention: antilock braking systems (ABS) and airbags. Neither are really new technologies, but their inclusion as standard items on American cars is new.

Antilock braking systems are more efficient than standard systems, because a vehicle can stop faster and remain steerable when the brakes lock and release many times per second, instead of locking up tight. Here is how ABS works: Sensors at each wheel detect how fast it is turning. If a wheel indicates that it is about to lock, the information is processed by an electronic control unit. The electronic control then directs the pump feeding pressure into the hydraulic unit to ease pressure on that wheel's brake. As the wheel speeds up, pressure is alternately increased and eased until the vehicle comes to a controlled stop. In effect, the ABS system rapidly pumps the brakes for you. Some 60 to 80 percent of a car's braking effort is accomplished by the front wheels. Because locked-up front wheels tend to go straight ahead regardless of which direction the wheels are turned, ABS represents a major safety breakthrough, allowing the front wheels to be steerable under heavy, emergency braking.

A revolutionary braking system still in the development stage uses a radar-sensing device to detect the presence of "threats" and activates an emergency braking system to decrease the impact speed rapidly when collision is imminent. This system, under development by RCA, is not intended to replace driver skill but rather to supplement it. It is intended to work only if something goes wrong—for instance, if the driver falls asleep at the wheel, has a heart attack, or is incapacitated for some other reason.

Both ABS and radar-actuated braking are examples of *feedback control systems.* In a feedback system, output information is continually sent back to the input of a machine or device to help control the overall operation. A familiar example of a feedback device is the thermostat that controls room temperature. A more complex example would be the automatic pilot that maintains aircraft in straight and level flight.

So far I have discussed what are called accident-avoidance systems, but what happens if you are unlucky enough to be involved in a car crash? Statistically speaking, every driver is involved in a serious accident once in every 10 years of driving, so give that eventuality a little thought.

A physical principle connected with accidents is called *conservation of momentum*, which states that the momentum of a system has constant magnitude and direction if the system is subjected to no external forces. The abbreviation g is a measure of Earth's gravitational pull on a body—thus a force of 2 g's is equal to twice the force of Earth's gravity. Translated to a two-car collision, this means that the g forces are roughly proportional to the relative weights of the colliding objects. In a collision between a full-size car (4000 pounds) and a compact car (2000 pounds), the g forces in the compact would be double those in the full size.

There is no doubt that driving small cars is hazardous to your health and well-being. Accident data show that occupants of small cars are 8 times more likely to be killed in a car crash than occupants of large cars and 10 times more likely to be injured. All of this leads to the subject of restraint systems, which are intended to keep the bodies inside the vehicle from continuing their forward momentum even when the car has come to a sudden and unexpected stop. The best possible restraint system today employs both a shoulder harness/seat belt and an air bag.

An air bag is one type of automatic crash protection equipment now available on many new cars. It is an extraordinarily effective safety device built into the steering wheel or dashboard. In a serious frontal crash—equivalent to hitting a brick wall at a speed greater than 12 miles per hour—crash sensors activate the air bag. Within 1/25 second after impact, the bag is inflated to create a protective cushion between the occupant and the vehicle steering wheel, dashboard, or windshield. The air bag then deflates rapidly. Nitrogen, which composes 78 percent of the air we breathe, is the gas that inflates the bag. A solid chemical, sodium azide, initiates the action that generates the nitrogen to inflate the bag. Considerably over one half of the fleet of new cars sold in the U.S. will have air bags in the early 1990s.

SUPERSONIC AIR TRANSPORT

Our current fleet of passenger jets is aging, and new planes will be needed soon. At present we fly to our destinations in subsonic aircraft (except for those few who are privileged to fly in the supersonic Concorde) powered by conventional jet engines and controlled, for the most part, by human pilots.

All this is on the verge of drastic change. Speed, power, and control are the three areas of expected technological change. A quick look at the projected advancements is in order for the would-be technologically literate.

Speed

New-generation passenger planes are on the drawing board today, and they all involve significant increases in speed. Today's subsonic aircraft fly at velocities under the speed of sound, which, as I have said, is 740 miles per hour at sea level. The British/French Concorde is a supersonic, and flies at Mach 2, or twice the speed of sound. A Mach number, named after the Austrian physicist Ernst Mach, measures the ratio of an object's speed to the speed of sound. When the Mach number exceeds 1, the object is moving at supersonic speed.

High Speed Civil Transport

The first of the new age high-tech passenger aircraft, called the High Speed Civil Transport (HSCT), is currently the focus of attention at NASA, Boeing, and McDonnell Douglas. The HSCT will be capable of flying at speeds just under Mach 3. At this speed a plane could fly from Los Angeles to Tokyo in about four hours. By way of comparison, the Boeing 747 makes the trip in about 10 hours.

The proposed HSCT will have a range of about 6500 miles (compared with 3200 for the Concorde) and will seat 300 passengers—200 more than the Concorde.

Questions about noise and possible environmental damage will be raised about the proposed new plane, just as they were about the Concorde and the never-built U.S. SST of the early 1970s. The relatively few flights that today's 16 Concorde planes make each day are not considered a threat to the environment, but a large fleet—say,

1000—of the new jets could damage the ozone layer with the nitrogen oxides that it would dump into the atmosphere. Supersonic aircraft cruise at around 60,000 feet, which is about twice the altitude of conventional airplanes. As you saw in Chapter 4, pollutants dumped into the stratosphere accumulate year after year, because the stratosphere does not clean itself as readily as does the lower level troposphere, and there seems to be some correlation between nitrogen oxides and ozone depletion. Minimizing exhaust effluents could prove to be the number one technical challenge facing the HSCT aircraft designers.

National Aero-Space Plane

Also called the X-30, the National Aero-Space Plane (NASP) is now in early development by the Air Force and NASA. Designers envision the day when this hydrogen-burning space plane will take off from an ordinary runway and then accelerate in a steep climb into the high stratosphere. It is designed to reach an ultimate velocity of 25 times the speed of sound as it passes near the edge of space 50 miles up.

The technology to build a Mach 25 plane is not available at this time. For instance, temperatures of the plane's skin will reach 5000 degrees Fahrenheit (2750 degrees Centigrade). No current material could stand it. Perhaps 60 percent of the proposed plane's takeoff weight will be liquid hydrogen, forcing engineers to find new materials that are light enough, strong enough, and heat resistant enough to do the job. Although research and development work is underway, aerospace engineers say that it will take at least two decades to solve the X-30's formidable engineering problems. In contrast, the HSCT could fly soon after the year 2000.

The X-30 will be developed no faster than the propulsion technology, and it is in this area that most of the current design efforts are concentrated.

Power

Advancements in jet propulsion are necessary to future high-speed air travel. Most of the planes in which we travel today are powered by jet engines. They are what keeps us up there while we watch the in-flight movie or have a meal, so you should have some idea about how they work.

As Figure 8-4 shows, the driving force in a jet engine is the same one that makes a toy balloon shoot forward when its opening is released and air escapes. We have stumbled across another physical principle here. It is called Newton's Third Law of Motion: "For every action there is an equal and opposite reaction." In the case of a jet engine, the motion of the expanding escaping air in one direction results in an equal motion, or thrust, in the opposite direction. In a jet engine, the burning of fuel and air produces hot gases that surge out the exit nozzle, producing thrust that drives the plane forward.

The difference between a jet engine and a rocket engine, which also ejects gas to produce thrust, is that the rocket engine does not breathe air—it carries all its own fuel, including an oxidizer, usually in the form of liquid oxygen.

Turbojets, which power most of today's aircraft, use an air compressor in the inlet to compress the incoming air, which is mixed with fuel and ignited in the combustion chamber. The expanding exhaust gases spin a turbine that both produces thrust and powers the compressor. Turbojets, however, have their limitations. As air rams into a turbojet engine, it encounters drag and consequently heats up. At speeds of about Mach 3, the combined effects of aerodynamic heating and combustion raise the internal temperature of the gases that turn the turbine to a point that is beyond the ability of current state-of-the-art materials and cooling techniques to handle. Thus, Mach 3 is the practical limit for conventional turbojet engines.

Ramjets are the simplest of all the jet engines in that they do not use turbines or compressors but rather rely on forward motion to ram air into their combustors. The air, heated and compressed by the ramming, then mixes with fuel and burns. Ramjets have to be brought up to speed by some other form of propulsion, because the air ramming effect does not take place at speeds below Mach 1.5. The upper limit for ramjets is about Mach 6, where a decrease in fuel efficiency takes place. At Mach 6 molecules of fuel and air break up into unburned and partly burned particles. Combination turbojet/ramjets have been considered, with the turbojet handling takeoffs and landings and the ramjet used during hypersonic cruising.

NASA is currently wind-tunnel testing a supersonic-combustion ramjet engine called a *scramjet.* It is believed that the scramjet could reach speeds of Mach 12. Defense Department studies suggest that the

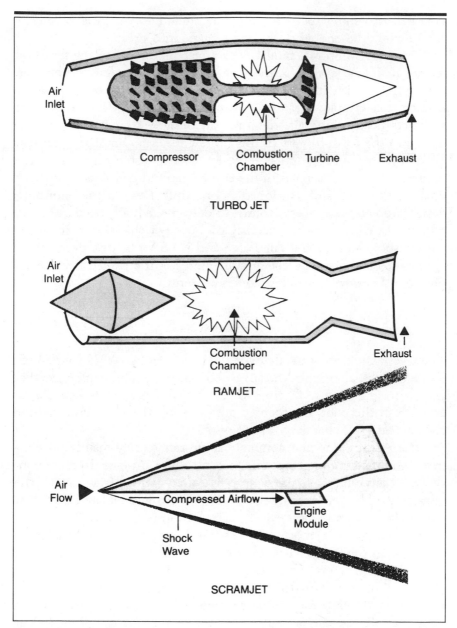

Air
Inlet

Compressor   Combustion   Turbine   Exhaust
              Chamber

TURBO JET

Air
Inlet

Combustion   Exhaust
Chamber

RAMJET

Air
Flow   Compressed Airflow   Engine
                              Module

Shock
Wave

SCRAMJET

Figure 8–4 *Jet Engine Technology*

potential maximum speed for scramjets may be as high as Mach 25, which would be enough to propel a plane into orbit.

Like a conventional ramjet, a scramjet relies on a plane's motion to ram air into the front opening at high speeds. Scramjets, however, are designed with internal airflow that prevents the extreme over-heating that cripples a ramjet traveling faster than Mach 6. Because scramjets need a long air inlet, aircraft designers must integrate the inlet into the forward part of the aircraft fuselage. Similarly, the rear portion of the fuselage must serve as the exhaust nozzle. Thus, the entire aircraft becomes part of the engine—an integrated assembly that must be designed and developed as one unit. Despite the enormous design problems, scramjets promise to deliver the high-speed capability of a rocket with the fuel efficiency of a jet. For this reason scramjets are the subject of intense Air Force and NASA research programs.

As speeds of commercial aircraft increase, the need for more so-phisticated control systems becomes evident.

Control

The recent addition of the fly-by-wire French Airbus A320 to the U.S. domestic fleet, along with the advanced generation 747 jumbo jets that can be programmed to fly from California to Tokyo with virtually no effort from the pilot, has focused attention on the redesign of airline cockpits and the changing role of pilots.

The A320 is the first commercial aircraft to rely mainly on elec-tronic commands and computers to control the airplane. In more con-ventional aircraft, pilots use a large wheel or control stick, which the pilot grasps like the steering wheel in a car, to manipulate cables that physically start and stop devices such as motors or hydraulic pumps. These devices in turn move the mechanical systems such as the ailerons and flaps on the wings, elevators on the horizontal surface of the tail, and the rudder, all of which control the flight.

The pilot in the A320 flies the plane with a controller that is similar to the joystick used for computer games. This device is known as a "side stick" because it is mounted at the pilot's side to permit a clear view of the display panels. Movement of this controller sends electrical signals to computers that translate them into commands to the aircraft's control surfaces.

This approach to aircraft control is called "fly-by-wire" and has been used for years in high-performance military aircraft. The main advantage of the system is that the electrical network is ideally suited for more extensive use of computers. Planes under computer control can respond more quickly to turbulence and other changes in flying conditions.

Computers can also be programmed to prevent the pilot from forcing the plane into a maneuver it was not designed to handle, such as a stall or a too sharp turn. This latter feature is called the "envelope protection system" and it is the subject of controversy among pilots and aircraft design engineers. Some experts want the pilots to have the option of extreme maneuvers in an emergency situation, but others point to those situations where the pilots have flown out of the design envelope and gotten into serious trouble. This aspect of fly-by-wire remains controversial.

A second evolutionary change in commercial aircraft involves the means by which information is transmitted to the pilot. In new aircraft, such as the Boeing 767 and 757 and the McDonnell Douglas MD-88, computers monitor aircraft systems, reporting on their status only if requested by the pilot or if something is wrong. In older planes the pilots have to monitor dozens of gauges and dials constantly to check engine, hydraulic, and electrical systems.

The new concept is called the "glass cockpit." In this system, information like airspeed, compass heading, altitude, and the like are monitored by computer and displayed on a screen when necessary. In effect, the computers tell the pilot how the computers are doing flying the airplane. The pilot is now free to do other things such as talk to the air controllers. As the cockpit becomes more automated, some airlines have reduced cockpit crews from three to two.

Despite fly-by-wire and the glass cockpit, pilots still maintain overall responsibility for piloting the aircraft, and they can turn parts of the systems off or on, flying the jet themselves as much or as little as they want.

If you travel by air very much these days you will appreciate all the new technologies, but you will also be quite aware that the major air-travel arteries in the United States, Europe, and Japan are clogging up so badly so fast that other alternatives may have to be considered. High-speed trains may be one option for the future.

HIGH-SPEED TRAINS

The fastest commercial train in the world today is France's TGV, which hits a top speed of 186 mph. Fast as it is, the TGV (for *train à grande vitesse* or train of great speed) is slow compared to Germany's and Japan's train, the experimental MAGLEV, a contraction of magnetic levitation. Both the German Transrapid and the Japanese MLU002 are designed for top speeds of between 250 and 300 mph.

Chapter 6 briefly discussed levitating trains (in connection with superconducting magnets), but the West German Transrapid is a different kind of magnetic levitating train that does not rely on superconductivity. This technology may be closer to commercial applicability. In fact the Transrapid is the leading candidate for a proposed Southern California to Las Vegas project. The MAGLEV would cover the 280-mile stretch to the gambling mecca in about 75 minutes as compared to 5 hours by car. (Think of it: You can go broke 3 hours and 45 minutes sooner!)

The German system uses conventional magnets rather than the superconducting ones used by the Japanese version. Another major difference in the two designs is the way the trains levitate. The West German model operates on the principle of attraction in magnetism and employs winglike flaps that extend beneath the train and fold under a T-shaped guideway. Electromagnets on board the Transrapid vehicle are attracted to a nonenergized magnetic surface. The interaction between the train's electromagnets and those built into the top of the T-shaped track lifts the vehicle approximately ⅜ inch off the guideway. This separation is controlled by an onboard automatic system. Another set of magnets positioned along the sides of the rail provides lateral guidance. In contrast, the Japanese model rests on a set of wheels until it reaches a speed of 100 mph, and then it levitates to a point 4 inches above the guideway.

Figure 8–5 shows a comparison between the Japanese repulsion system (described in Chapter 6) and the West German attraction system. Propulsion is basically similar in the two systems. In both cases the train effectively rides on an electromagnetic wave. Alternating current in a set of magnets in the guideway changes their polarity to alternately push and pull the train along. Raising the frequency of the current speeds up the train. Reversing the poles of the magnetic field reverses direction, which provides braking.

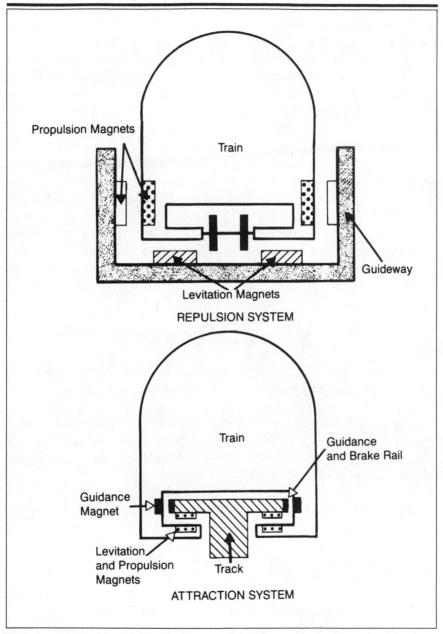

Figure 8–5 *Comparison Between Japanese Repulsion MAGLEV System and German Attraction System*

Because the Transrapid's winglike flaps wrap around the rail, it cannot be derailed. The Transrapid also has the advantage of needing much less space than a conventional rail track or highway.

Advocates of the MAGLEV technology believe that the United States should enter the competition. Senator Daniel Moynihan of New York has introduced bills that call for research, development, and construction of a U.S. MAGLEV system. Moynihan and others recognize the clear need for fast, efficient surface transportation for the shorter distance runs. Air traffic has doubled in the last 15 years and is expected to double again in the next 15. As airports become more overcrowded, delays are commonplace. Much of this congestion is caused by travel between cities less than 500 miles apart—the ideal distance for a 300 mph MAGLEV.

Gridlock from automobile congestion is another concern. Americans already own more than 125 million cars, and that number is expected to triple in 50 years. It may be that a fast, efficient train system could eventually supplement cars.

KEY CONCEPTS

▶ The First Law of Thermodynamics simply states that energy is neither created nor destroyed; it just changes form.

▶ The Second Law of Thermodynamics precludes perpetual motion machines, because it states that every time we convert one form of energy into another form, we lose energy (usually in the form of heat) on the exchange.

▶ Most of the cars driven today have reciprocating internal combustion engines using the Otto, or four-stroke, cycle.

▶ A Mach number measures the ratio of an object's speed to the speed of sound at sea level (740 miles per hour).

▶ In a two-car collision, the g forces (a measure of the Earth's gravitational pull) on the occupants are proportional to the weights of the colliding objects. If you are going to be in an automobile accident, arrange to be in the heavier car.

▶ Jet engines operate on a physical principle called Newton's Third Law of Motion: For every action there is an equal and opposite reaction. As hot expanding gases escape in one direction, thrust is exerted in the opposite direction.

▶ Magnetic levitating trains do not necessarily rely on superconductivity and, in fact, the first MAGLEV train in this country will probably use conventional magnets.

—9—

SUPERWEAPONS
and
ARMS CONTROL
TECHNOLOGY

Whater, I asked myself, is the least that an intelligent, technologically literate person should know about the difficult subjects of nuclear arms, superweapons, and arms control? I say *least* because these are not subjects to which we ordinarily give much thought. The nuclear sword of Damocles has been hanging over our heads so long we have become numb to the terror. We have deferred thinking about the unthinkable to the experts—to the military, the weapons industry, and the so-called defense intellectuals. It's just possible that "leaving it to the experts" was not the best decision we could have made in the last few decades. Apathy, it turns out, carried a huge price tag.

The opposite of apathy is involvement and the entrance fee for involvement is technical knowledge. Not that they make it easy. First there is what McGeorge Bundy calls "Nuclear Theology," the dark and arcane art of formulating a strategy for deterring nuclear war or, if that fails, fighting such a conflict. To communicate in this cloistered world we must understand the lexicon—the technical language including *deterrence, counterforce, flexible response,* and even *missile gap,* and *window of vulnerability.* Second there are the technical esoterics of the arms themselves: B-1s and B-2s, MIRVed and unMIRVed missiles, MX and Midgetman, SSBNs, SSNs, ASW, and SDI. Finally, consider the veritable thicket of arms control acronyms through which we must hack our way: SALT I and II, ABM, INF, and START.

Patience will be rewarded, however. Students who complete this short course will be more than able to hold their own in any parlor arms debate. They will, in fact, qualify as defense intellectuals, apprentice class. The diligent reader may, in fact, be able to play an active citizen's role in the arms game rather than merely standing on the sidelines in docile apprehension. If you get discouraged, remember the alternative is apathy, and we already know what that got us.

This chapter is organized according to the following topics:

- The lexicon of nuclear strategy

- The strategic weapons arsenal

- Star Wars—The Strategic Defense Initiative

- Nonnuclear superweapons

- The technology of arms control and verification

Weapons technology strongly influences policy, but before you turn to the weapons arsenal, take a moment to digest a few of the terms necessary to understand the nuclear arms debate.

THE NUCLEAR STRATEGY LEXICON

Deterrence: the military capability to discourage a nuclear attack by virtue of the capacity or threat of retaliating. This is the key concept of the nuclear age. Both powers—the United States and the Soviet Union—have the ability to destroy each other and neither side has a defense system that could be relied upon. Each side holds the other side's population in hostage and neither adversary knows how to escape from the balance of terror. Deterrence has been likened to a game of tick-tack-toe; if both sides always act logically, the game always ends in a draw. A draw is boring but it is better than a nuclear war.

Mutually assured destruction (MAD): sometimes called the balance of terror, MAD is the stated national policy arising from deterrence. In event of a nuclear attack, either of the two superpowers could absorb a nuclear first strike and still be capable of retaliation, causing an unacceptable level of damage to the attacker. *Unacceptable level* has been quantified in Pentagon documents as meaning one-third of the population and one-third of the industrial capability of a nation.

Triad: the air-based, sea-based, and land-based nuclear delivery systems that make up the three legs of the U.S. strategic weapons arsenal.

Counterforce: the employment of nuclear weapons against the weaponry and command and control centers of an adversary. A counterforce targets nuclear forces on an adversary's military capability to reduce the potential damage that could be caused by these weapons. This is also called damage limitation. Targets in this strategy are called "hard" because they are dug in underground or reinforced with concrete and steel.

Countervalue: the targeting of nuclear forces against an adversary's population and economic centers (that is, cities). This is the "hostage" strategy essential to deterrence. Countervalue targets are called "soft," in contrast to the hardened silos or command centers targeted under the counterforce strategy.

First strike capability: the ability to conduct a first strike so devastating that one's opponent is incapable of retaliation. Neither the United States nor the Soviet Union has ever had a first strike capability.

Flexible response: the ability to respond with measures in keeping with the provocation, that is, something less than massive, all-out retaliation. This strategy is a reality of today's weapons arsenal; arms can be launched selectively or massively against a repertoire of targets.

Decapitation: the targeting of nuclear forces against an adversary's leadership and command structure to limit the adversary's retaliatory capability. The target of a decapitation strike would be something called C³I.

C³I: Command, Control, Communication, and Intelligence; the people, equipment, and procedures used by the national leadership to monitor and direct the operation of military forces.

Launch on warning: the doctrine that calls for launching strategic forces upon receipt of information from a number of sensors (such as surveillance satellites and other means) that an adversary has initiated a nuclear attack. This is sometimes called the "hair trigger" response and is logically necessary only when there is concern about the vulnerability of retaliatory forces.

Launch under attack: a variation on the launch on warning doctrine that calls for launch of nuclear forces only after confirmation that enemy weapons have detonated on home territory.

Minimum deterrence: a policy of retaining only enough nuclear weapons to provide an assured destruction capability. This strategy does not provide for limiting damage or targeting counterforces.

Multiple independently targetable reentry vehicle (MIRV): two or more reentry vehicles or warheads carried by one ballistic missile, each of which can be individually directed toward a separate target.

Missile gap: political rhetoric with little or no basis in fact. One can make either side look as if it is ahead by the use of mathematical sleight-of-hand. For instance, one can make the United States look as if it is ahead in the arms race by counting total numbers of warheads, or one can make the Soviets look as if they are ahead by counting only Intercontinental Ballistic Missiles (ICBM) warheads and yield. Figure 9–1 shows the current U.S./USSR strategic weapons arsenal.

Yield: the destructive energy released by nuclear weapons, measured in megatons (millions of tons of TNT equivalent). The destructive

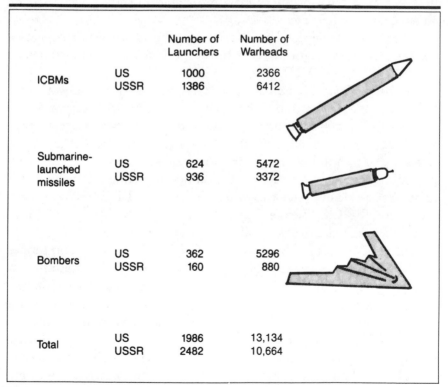

		Number of Launchers	Number of Warheads
ICBMs	US	1000	2366
	USSR	1386	6412
Submarine-launched missiles	US	624	5472
	USSR	936	3372
Bombers	US	362	5296
	USSR	160	880
Total	US	1986	13,134
	USSR	2482	10,664

Figure 9–1 *U.S.–USSR Strategic Weapons Arsenal*

power of modern nuclear weapons can be put into perspective if you consider that the sum total of munitions expended in World War II from 1939 to 1945 added up to less than 6 megatons. The total from that war is less than the megatonnage of only *one* of the large nuclear warheads now deployed. The relatively "small" bombs that devastated Hiroshima and Nagasaki had yields of a little more than one-hundredth of a megaton.

The total destructive power of the combined U.S. and USSR arsenals is a complex calculation based on a number of assumptions, but we can simplify the operation with a little back-of-the-envelope mathematics. If we assume a worldwide arsenal of 20,000 deliverable weapons (there are in fact many more, but let's be conservative), each capable of killing 1 million people, the total is a staggering 20 billion potential casualties or four times as many men, women, and children as there are on Earth. The word "overkill" is not one with which professional defense planners are comfortable, but it is difficult to find an adequate euphemism.

Backing off from the arms race has not been easy, but the process has been initiated and our basic defense-speak vocabulary must include SALT, INF, and START.

Strategic Arms Limitation Talks (SALT): a series of negotiations between the Americans and the Soviets that deal with offensive and defensive strategic arms. SALT I resulted in two formal agreements: the Antiballistic Missile (ABM) Treaty and an Interim Agreement, freezing the total number of long-range ballistic missiles. SALT II replaces the Interim Agreement with a long-term comprehensive treaty limiting strategic offensive weapons. Although SALT II has never been ratified by the U.S. Senate, both the Soviet Union and the United States have stated that they will not violate this treaty.

Intermediate-range Nuclear Forces (INF) Treaty: signed and ratified in 1988, this is the treaty that eliminated short- and medium-range nuclear missiles (with ranges between 300 and 3300 miles (500 and 5500 kilometers). Verification was provided for by including on-site inspection by both sides. The INF treaty is hailed as an important first step, but in fact it has had only a small quantitative effect on the nuclear threat. INF eliminated four times as many Soviet warheads as American. But the number of weapons eliminated by both sides was less than 5 percent of the total number deployed.

Strategic Arms Reduction Talks (START): current negotiations over reductions in strategic offensive weapons. The numerical limits have been agreed upon—each side will be restricted to no more than 4900 ballistic missile warheads and a combined total of 6000 "accountable weapons." Strategic nuclear delivery vehicles (SNDVs), the sum of missile launchers, bombers, and cruise missile carriers, will be limited to 1600. The details are complex, but it appears that each side would in reality have 8000 to 9000 deployed weapons, only a 25 to 30 percent reduction from current levels, not the 50 percent commonly heard.

With the help of these buzzwords, you are well equipped to carry on parlor arms control negotiations. If, however, we are to be part of an informed and effective public constituency we must understand more about the arms themselves—both the nuclear arsenal and the nonnuclear superweapons.

THE STRATEGIC WEAPONS ARSENAL

The U.S. strategic delivery systems are divided into what is called the triad: air-based, sea-based, and land-based systems. To ensure a reliable retaliatory force, it was not considered wise to place all our reliance on any one weapons system, and we didn't. Each element of the triad has its own operating characteristics. An alert strategic bomber force can, given enough warning, scramble and escape damage. The nuclear ballistic missile submarines move virtually undetectable under the ocean and therefore cannot be taken out with a first strike. The land-based ICBMs are loaded into hardened underground silos that, until recently, were considered invulnerable to attack. These forces provided the United States with a secure deterrent. The triad concept has been considered sacrosanct and untouchable for over 40 years, but technical realities are now beginning to encroach on it.

The highly secret war plan that ties all these forces together is called the *Single Integrated Operational Plan (SIOP)*, and it was revised again this year to take into account the technical characteristics of the MX intercontinental missile and the B-1 bomber. The SIOP provides the president with dozens of options under which to employ the nation's 13,000 strategic nuclear warheads in an emergency. Options range from an all-out response, through a protracted nuclear war, to a controlled "small" barrage response. I once had the need to read the

SIOP (under the watchful eyes of an armed Marine guard), and it is not an experience I recommend. Its bland phrasing (such as "minimum acceptable megadeaths") will, as Flannery O'Connor once said about the pain of creative writing, "make your hair turn gray and your teeth fall out."

Now let's see what we get, in part at least, for the 26 cents out of every tax dollar earmarked for defense.

Air-Based Strategic Systems

The B-1 Bomber

Air-based strategic weapons systems are the weakest link in the triad— so weak that they may not in fact exist at all. Few weapons in the U.S. arsenal have been as controversial as the B-1 bomber, now called the B-1B (replacement for the aging B-52). Even though the B-1 has been built and paid for, the aircraft continues to be the subject of heated partisan contention. Its implementation was originated by President Nixon, "killed" by President Carter, and resurrected by President Reagan.

The dispute over the B-1B bomber has continued, even though most of the $28 billion for the program has been spent and all 100 of the planes delivered (two crashed in 1988, leaving a fleet of 98). At the time of its delivery to the Air Force by its manufacturer, Rockwell International, the B-1B was the most expensive airplane in aviation history.

The only problem is that there is now considerable doubt about the B-1B's capability to carry out its mission of a nuclear strike against the Soviet Union. Oh, the born-again bomber, as it is sometimes called, flies adequately and would have been a great weapons system in World War II before air defenses became sophisticated. The questions concern the ability of the outdated B-1B to penetrate the modern Soviet defense systems in significant numbers. If it can't do that, it is not a viable element of the strategic triad.

Even the Air Force admits that the B-1B bomber fleet is not fully "mission capable," which in Air Force parlance means the ability to carry out an assigned task. In the case of the B-1B, that would mean a nuclear strike against the Soviet Union.

At issue are the bomber's defensive electronics, which are intended to detect and jam enemy aircraft and missile radars. Because attacking aircraft and missiles would use radar for targeting the B-1B, it is essential that the bomber be capable of jamming or otherwise nullifying enemy radar. These electronic countermeasures include more than 100 antennas and devices intended to simultaneously detect many radar signals from fighter aircraft, ground stations, and missiles fired from the ground or from interceptor planes. Much of this system is secret, but the Air Force has stated that the ability of the system to handle the entire spectrum of threats simultaneously is deficient.

Representative Les Aspin, Chairman of the House Armed Services Committee, has been one of the more outspoken critics of the B-1B. He contends that if you run this bomber against theoretical Soviet antiair defenses, you have a "serious shortfall in capability." Specifically, according to critics, the bomber is designed to attack at high subsonic speed while flying only 200 feet above the ground. Thus the bomber can foil enemy fighters shooting down at it but not surface-to-air missiles shooting up. As a result, Representative Aspin claims the bomber will never do what it was advertised to do and has said that he "doubted Congress would have approved the B-1B program if we knew it didn't have workable electronic countermeasures."

It takes more than a decade to develop a new high-tech weapons system, be it a new tank, jet fighter, or manned bomber, and once the Pentagon is committed to a multibillion-dollar weapons system, the program acquires an unstoppable momentum even if the weapon has become obsolete. Senate Armed Services Committee Chairman Sam Nunn cites the B-1B as "a key example of this kind of waste." Critics of the B-1B say that it is a beautiful aircraft and a fine example of high technology, just so long as the Air Force does not try to fly it where there is a war going on.

The problem now is what to do with the B-1B fleet. Over the next few years, Congress will have to decide whether to invest substantial additional monies in the B-1B bomber program to enhance the aircraft's ability to penetrate Soviet air defenses and to make it useful in a wider range of combat missions.

Another alternative may be to change the B-1B's mission from penetrator to a "shoot-and-then-penetrate" role in which it launches externally carried cruise missiles before penetrating Soviet defenses. Another alternative is to employ the B-1B as a "standoff" missile launcher, carrying only long-range cruise missiles and staying outside the range of defensive systems.

The B-2 (Stealth) Bomber

If you thought the B-1 bomber was expensive at $278 million per aircraft, what are we to think of the B-2 at a mind-boggling $500 million per aircraft? The Air Force has asked for $60 billion for a fleet of these radar-evading superbombers. To put that figure into some perspective, it is more than the U.S. Department of Agriculture's $55.5 billion budget and five times the National Aeronautics and Space Administration's $11.5 billion budget for 1988.

Critics of the Stealth bomber have focused on its cost, because there is not much else known about it. The Air Force has been deliberately vague, and sometimes contradictory, both about the Stealth's design capabilities and about its strategic mission.

Because manned bombers take several hours to reach their targets, they are not usually considered to be first-strike weapons. The B-2 must then be seen as contributing to our overall deterrent or our warfighting capability. The Air Force has said that in a nuclear war, the Stealth bomber's mission, after the Soviets had fired missiles and the United States had retaliated with missiles, would be to destroy unfired mobile missiles and the Soviet command center. Did we hear that correctly? After most of us are dead, the Stealth is going to sneak through what's left of Soviet air defenses and take out their command center!

In defense of the B-2, Air Force Chief of Staff General Larry Welch has said: "The key to deterring nuclear war is for the Soviets to never believe that they can succeed in achieving their objectives. And the way to do that is for them to clearly understand that all of those things they hold to be of great value are held at risk."

Several groups of U.S. scientists and arms control advocates have urged Congress to consider canceling the Stealth bomber. Despite its much heralded ability to evade enemy radar, they point out, the B-2 still would be susceptible to Soviet satellite detection, and its survival would be threatened by its slow speed, limited flying range, need for aerial refueling, and lack of weapons to defend itself.

With the B-2's more than $500 million per aircraft price tag, the Air Force may have finally gone too far. The Air Force's romantic attachment to manned bombers has been likened to the lure of excitement and tradition that the cavalry had for European armies at the turn of the century. Just as the technological innovation of the machine gun made cavalry charges obsolete, the missile age may have caught up with manned bombers.

The Stealth advocates and critics will have much to say on this subject before the final decision is made. In any case, the current plan to have 100 of these radar-evading planes in service by the end of the century is going to come under hard scrutiny.

Maybe we should leave the last word on the Stealth to comedian Bob Hope. "If it's supposed to be invisible," he said, "why don't we *not* build it and tell the Russians it's there?"

Sea-Based Strategic Systems

Ballistic Missile-Firing Nuclear Submarines (SSBNs)

Because of the increasing vulnerability of land-based ICBMs and doubts about the mission capability of manned bombers, both the United States and the Soviet Union rely on ballistic missile-firing submarines as the ace in the hole for deterring nuclear war. These SSBN systems could wait out a first strike against land-based strategic forces and still deliver a devastating retaliatory blow. At present, the United States has 36 ballistic missile-firing submarines, of which at least 25 are at sea at all times. Under the projected START agreement, the United States would possess 18 missile-carrying submarines. Of these, a maximum of 12 could be at sea at one time.

Submarine-Launched Ballistic Missiles (SLBMs)

The latest in the submarine-launched missile series is the Trident 2. The Trident 2 is a solid fuel, three-stage weapon with a range of nearly 6000 miles. Each Trident 2 is capable of delivering 10 highly accurate warheads to different targets. Each of the 20 or so Trident submarines will carry 24 of the Trident 2 missiles.

Some defense analysts have raised questions concerning the Trident 2 missile's accuracy. If the role of the SLBM is assured deterrence, these analysts ask, why does the weapon have to be so accurate? Presumably in a deterrence role, the targets would be cities (soft targets) and a high degree of accuracy is not required. Only experts with access to the top secret strategic planning document, the SIOP, can answer that question, but it may be a case of technological imperatives—that is, the technology was there to make the Trident 2 highly accurate, so why not do it?

Submarine-Launched Cruise Missile (SLCM)

The Submarine-Launched Cruise Missile (SLCM, pronounced *slickem*) is a 20-foot-long "Tomahawk" missile deployed on both submarines and surface ships. This SLCM is really a small, unmanned jet capable of flying a few hundred feet above the surface for distances of up to 1500 miles with a nuclear warhead, or about half that distance with a heavier conventional warhead.

Cruise missiles are nothing new. The "buzz bombs" that the Germans employed in World War II were cruise missiles and the United States has had a variety of cruise missiles in its arsenal over the years—Regulus, Matador, Mace, and Snark were the names of a few of them. Major problems with early cruise missiles involved their poor accuracy compared with ballistic missiles and their vulnerability to air defenses.

Two Tomahawk technical innovations have overcome these limitations. One innovation is the use of terrain-contour-matching (TERCOM) to allow in-flight updating of the Tomahawk's guidance system. TERCOM works like this: Stored inside the missile's computer is an array of digital maps displaying the contour of the Earth's surface at certain checkpoints along the missile's preprogrammed flight path. A radar altimeter points downward from the missile and precisely measures the missile's altitude by bouncing a radar beam off the ground. At specified points along the way, the TERCOM computer compares the readings with the appropriate contour map. If the missile is off track, the computer searches the map sector close to where it should be and finds a better match. Then, TERCOM steers the missile back on the right course. TERCOM can put the Tomahawk within a few feet of its target.

The second technical innovation is a terrain-following flight system that allows the missile to detect and avoid hillsides and other obstacles. This system allows the Tomahawk to avoid radar detection and sneak in to its target.

Altogether the Tomahawk is a formidable weapon, but therein lies the problem. From the point of view of the arms control community, the problem with SLCMs is their dual capability. It is virtually impossible to tell the difference between a nuclear and a conventionally armed SLCM. To count them or control them, as an arms control agreement would require, on-site verification measures would be necessary, and the Navy is none too happy about the thought of having Soviet observers on U.S. submarines.

National Security Advisor Brent Scowcroft, a leading player in START negotiations, has suggested that deployment of SLCMs may not be "advantageous" to the United States in the long run. The employment of SLCMs by either side complicates the problem of arms control enormously. Sidney Drell, for many years an advisor to the United States on national security and arms control issues, calls SLCMs "the major sticking point—the most difficult problem we have right now regarding START." He has also said, "It would be nice if there were no long-range SLCMs."

Carrier Groups

If the Air Force can be accused of being unrealistically romantic about its beloved manned bombers, the Navy may be in the same position over carrier groups. Carrier groups are described here because of their strategic role—that is, they have the capability of launching nuclear strikes. They do have a conventional war-fighting capability apart from their strategic role.

Carrier groups are floating cities centered around a massive aircraft carrier. A modern Nimitz-class nuclear-propelled ship carries some 90 airplanes—F-14 fighters, F/A-18 fighter/attack planes, A-6 and A-7 attack bombers (capable of carrying nuclear bombs), two early-warning aircraft, antisubmarine patrol aircraft, helicopters, and transport and utility planes.

The carrier is surrounded by a fleet of smaller combat and utility ships. Typically, these include two frigates (ships ranging in size between a destroyer and a cruiser and armed with guided missiles to help protect the carrier), two cruisers, and two destroyers. The combat ships are armed with a variety of cruise missiles, antiaircraft surface-to-air missiles (SAMs), and antisubmarine warfare gear. At least one of these support ships is an Aegis cruiser equipped with electronic systems designed to detect attacking planes or missiles and to battle-manage the carrier group's defense weapons systems. (The Aegis battle management capability is discussed later in this chapter.)

It's all quite impressive, and it's all quite expensive. A carrier group costs around $18 billion to build, employs more than 8000 people, and runs about $1 billion a year to operate. As impressive as this floating airport is, it carries little offensive striking power. Of the 90 or so aircraft on the carrier, only about 40 are attack vehicles. All the rest, and all the elaborate antisubmarine and antiaircraft systems, are there to protect the carrier. In other words, about 95 percent of the enormous cost of the carrier group is devoted to buying and defending a floating airport capable of mounting only a small attack. The cost effectiveness of this entire concept is likely to come under some painful (to the Navy) reconsideration in the budget-conscious 1990s.

Land-Based Strategic Systems

Intercontinental Ballistic Missiles (ICBMs)

Early in the evolution of the U.S. ICBM forces, the missiles were mounted on launch pads out in the open, but it soon became evident that in this configuration they were vulnerable to attack. The next step, then, was to place the missiles in hardened (concrete and steel) underground silos. Unfortunately, increased accuracy of Soviet missiles may have made even the hardened silos vulnerable to surprise attack.

The U.S. ground-based ICBM force in early 1989 consisted of 1000 Minuteman missiles located in fixed silos at seven bases in the North Central U.S. Under START, the United States is likely to retire our 450 aged single-warhead Minuteman II missiles. The remaining Minuteman III missiles have three warheads each.

America's answer to the vulnerability of our fixed land-based missiles was the MX. After much debate, Congress approved the construction of 50 MX missiles, each of which has 10 independently targeted warheads. The problem was what to do with the MX—specifically where and how to base it.

The MX has always been a political weapon—intended to close the "window of vulnerability" that only some defense analysts perceived as open to begin with. Basing schemes for the MX became even more political. President Carter's plan to scatter 200 MX launchers in 4600 shelters on special roadways in remote parts of Utah and Nevada—the shell-game approach—was scrubbed when President Reagan took over. Secretary of Defense Weinberger favored deploying the MX in large, specially designed jets—called the Big Bird approach. None of the many basing schemes stood up to technical scrutiny, and the Reagan administration fell back on placing them in the Minuteman silos. This compromise solution caused critics to point out that if the silos were vulnerable with Minuteman missiles, were they not just as vulnerable with MX? Despite the critics, the MX entered the Air Force's inventory as a silo-based weapon.

In 1989, then Defense Secretary Frank Carlucci authorized a full-scale development of a rail-car launching system for the MX. The incumbent Secretary of Defense Richard Cheney has concurred. As currently configured, the rail garrison system will consist of 50 MX missiles, based aboard 25 trains, with each carrying two missiles. Eventually, all 100 MXs would be placed on trains, according to current plans.

An alternative solution to the silo vulnerability problem involves the construction of 300 to 500 single-warhead missiles given the name Midgetman. The idea is that these small missiles can be scattered over a large area and thus present a problem to a potential attacker.

The ICBM controversy of the moment is whether the single-warhead Midgetman or the multiple-warhead, railroad-based MX can best survive a surprise attack and thus provide a more believable deterrent. The Air Force wants the MX, and many in Congress want the Midgetman. Based on past experience, the taxpayer is likely to be asked to pay for both. And the real question may be whether either is necessary.

The strategic arsenal reviewed here provides an overview of U.S. offensive weapons. Strategic defense is a different game and requires a different perspective. That brings us to Star Wars.

WHATEVER HAPPENED TO STAR WARS?

When President Ronald Reagan first proposed the Strategic Defense Initiative in a nationally televised speech on March 23, 1983, he dramatically shifted the focus from offensive missiles to a space-based defense against them. What Reagan and a small group of advisors dreamed about, it appeared, was a sort of giant Astrodome over the continental United States. Attacking missiles would be destroyed by lasers, particle beams, and other "directed energy" weapons deployed in space. SDI would, Reagan told us, "render nuclear weapons impotent and obsolete." Star Wars, as it came to be called, was and is an enormously appealing fantasy—a defense so perfect that not only would Soviet nuclear weapons be rendered useless, but ours would then be unnecessary. "Would it not be better," the president asked, "to save lives than to avenge them?"

In a later speech Reagan described SDI as "a shield that could protect us from nuclear missiles just as a roof protects a family from the rain." Animated television spots conveyed this concept to the American public. Some experts referred to SDI as a moat in space, others as a space Maginot line, and the experts implied that SDI could be as easily outflanked. In any case, to the amazement of our allies and adversaries alike, nuclear deterrence—the policy that had anchored peace for decades and defined relationships between big governments— was out, and a new defense of some kind was in.

Defense of what kind? The details came along a little later. Here be sure you understand the differences among (1) SDI as originally defined; (2) the modified SDI that came along a year or so later after technological reality set in; (3) the considerably revised SDI plan of October 1988; and (4) the Bush Administration concept, which is Star Wars in name only. If you are going to debate the pros and cons of SDI, and everybody does, you should define which SDI you are talking about.

Original SDI

The original vision (or dream, depending on which side you took in the great debate) involved four important components:

- a space-based surveillance, tracking, and targeting system that could detect a launch immediately and use a coordinated combination of radar, lasers, and infrared sensors to guide U.S. missile-destroying weapons

- an arsenal of directed-energy weapons of two types: lasers and "particle beams" intended to destroy enemy missiles as they leave their launchers and before the warheads separate from the guided missile (that is, between the boost and postboost phases). See Figure 9-2 for the four phases of ICBM flight

- a family of kinetic-energy weapons that carry no explosives but rather are intended to destroy their targets by means of their energy of motion alone. Called "smart rocks," these weapons are intended to intercept incoming missiles in the midcourse and reentry phases

- a battle-management system using a supercomputer to coordinate every element of the defense

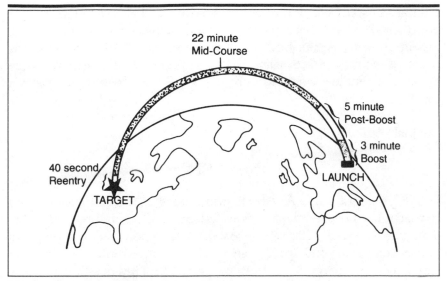

Figure 9–2 *Four Phases of ICBM Flight*

Grand and imaginative as the original concept was, countermeasure options were always open to the Soviets to nullify this system. They could fly under the SDI system with cruise missiles, overwhelm it with numbers, outsmart it with decoys, or outfox it by blinding the necessary sensors. Each of the countermeasures would cost far less than would Star Wars.

A word is in order here on "directed-energy" weapons—the heart of the original SDI. Four kinds of lasers were considered as kill weapons during the boost phase: chemical, excimer, free-electron, and X-ray. The beams produced by all of the lasers travel at the speed of light. Any target could be hit practically instantaneously, but would the amount of energy directed to a target be enough to produce a "kill"?

A panel of scientists from the American Physical Society reviewed the technical feasibility of each of these systems and concluded that the power requirements for SDI weapon use of these laser systems far exceeded any current technology. Several decades of intensive research would be necessary, the American Physical Society stated, before a final conclusion could be reached on the technical capability of these proposed antimissile laser weapons.

The other type of directed-energy weapon considered for SDI consists of particle beams. A particle beam is produced by first accelerating a charged beam of negative ions (atoms that have an additional electron). The extra electron is then stripped away in a gas cell, leaving a neutral beam. Again, however, the American Physical Society was negative about the practical application of particle beam weapons. Voltage and current levels needed were at least 20 times higher than today's capability.

In short, technological reality had set in. A modification to the original plan was in order.

Modified SDI

In 1987, General James A. Abrahamson, head of the SDI organization at that time, unveiled what was called Stage One architecture. This plan consisted of 3000 space-based, rocket-powered interceptors housed on about 300 space platforms. The interceptors would use *kinetic energy* (energy of motion) to destroy enemy missiles. Kinetic-energy weapons are sometimes called "smart rocks" and they are intended to destroy enemy missiles through impact at high speeds.

The modified SDI still included a series of surveillance and tracking satellites, with an elaborate command, control, and communications system also based in space. As a backstop to all the space-based anti-missile systems, ground-based rockets would be used to intercept incoming Soviet warheads in the reentry phase. The major emphasis in the Stage One plan was on interception or kill from space while attacking missiles were in their boost phase. This plan was widely criticized for attempting to rush into deployment with today's rather than tomorrow's technology. The exotic directed-energy systems were postponed and penciled in for Stage Two. A Senate study showed that, even if Stage One met all its requirements, it would stop only 16 percent of incoming warheads in a full-scale attack. The cost of Stage One was estimated at $115 billion and, in a era of budgetary restraint, this meant that SDI might jeopardize other weapons procurement. The Joint Chiefs of Staff concluded that Stage One had to be revised.

Revised SDI

In the revised plan of October 1988, the system's costs were scaled back to $69 billion. More than half of the space-based interceptors were discarded and the quality of the space sensors was reduced. Battle management responsibility was returned to Earth. In the revised plan, ground-based interceptors were increased by 70 percent. Deployment was scheduled for the year 2000. In fact, SDI had gone through a drastic metamorphosis. A high-tech, ground-based, antiballistic missile system it was; Star Wars it was not.

SDI suffered still another blow in January 1989 when the prestigious National Research Council's high-level committee on advanced space-based power technology released its findings. The committee members stated that the technology to produce energy needed for the planned Star War battle stations could take a half-century to achieve.

Various types of advanced nuclear reactors for powering space weapons could weigh up to 3205 tons, the NRC report found, and each would have to generate as much as 500 million watts of electricity—equal to the output of many large commercial nuclear plants on Earth. The Space Shuttle, currently the United States' most powerful launch vehicle, can lift only about 27 tons into orbit.

The Bush Administration Star Wars and Brilliant Pebbles

Instead of groups of large satellites capable of launching 10 rockets each, as planned under President Reagan, the Bush Administration plan calls for the deployment of thousands of small individual rockets in orbit. Dubbed "brilliant pebbles" because they seem to be an advancement over the smart rocks called for under previous Star Wars concepts, these space-based weapons would track enemy missiles and pick targets, eliminating much of the need for outside guidance from huge sensor satellites and ground stations.

Each weapon would be about three feet long and weigh about 100 pounds. It would have its own computer brain and an innovative wide-angle optical sensor to detect the hot exhaust of attacking missiles and to pick out targets. Each rocket would have its own guidance system and would be capable of operating without control from the ground.

The idea is to sow space with 10,000 to 100,000 of these small weapons that would home in on enemy missiles and destroy them by force of impact. Critics, however, say this plan is seriously flawed and, like the old Star Wars concept, easy to counter. Roy E. Kidder, a Lawrence Livermore Laboratory physicist who is critical of Star Wars, said brilliant pebbles might be effective against the current generation of SS-18 missiles, which fire their engines for 6 minutes, but it couldn't possibly work against a fast-burn booster that fired its engines for 60 seconds or less.

The Bush Administration approach also calls for slowing the pace of the Reagan program by delaying the development of a new tracking satellite, a ground-based interceptor rocket, and an orbiting laser. The development of these systems could violate the 1972 Antiballistic Missile Treaty.

In addition to technical changes, the new approach involves changes in SDI goals. Defense Secretary Richard Cheney has said that the original Star Wars concept was "oversold" in the Reagan years, and Pentagon officials now promote it as a way of enhancing the nuclear deterrent rather than rendering missiles obsolete or providing an impermeable shield.

The stated goal of the 1988 SDI concept system was to intercept half the warheads from the Soviet SS-18 missiles. Under START, the USSR agrees to eliminate this same number of SS-18s. If the Soviet agrees to do for free what the administration proposes to spend billions to do, SDI may be in for some hard times budgetwise. Nervous about this criticism, the current head of the SDI office, Lieutenant General George L. Monahan, Jr., has said that the antimissile system would be able to eliminate a portion of a Soviet nuclear attack, but the exact percentage is classified.

NONNUCLEAR SUPERWEAPONS

Has Technology Forced People Out of the Loop?

The brilliant pebbles concept for Star Wars is an example of new weapons technology that almost seems to take humans out of the defense system loop. This is a subject of growing concern to many. Warfare in which machines make decisions and soldiers or sailors fire on unseen enemies may be the curse of high-tech weaponry. Improved weapons technology has led to a clear imbalance between the ability to kill at long range and the ability to see who is being killed. The speed, range, and accuracy of modern weaponry now means that there is almost no time left to verify the targets. Two tragic examples of the imbalance between killing technology and sensor technology make this point.

The Iraqi pilot who sent two Exocet missiles crashing into the U.S. destroyer Stark in 1987 almost certainly did not want to damage an American ship. He probably saw a blip on his radar screen and fired on the information his sensor equipment provided. He could have flown closer to his target in an attempt to identify its nationality, but to do so would have meant risking his own plane and, of course, his own life.

In the same way, Captain Will Rogers of the USS Vincennes had little choice but to fire the missiles that destroyed Iran Air Flight 655 in 1988, killing the 290 civilians on board. Captain Rogers could not wait to identify the incoming aircraft visually, because to do so would have placed his ship within range of a possibly hostile aircraft's missiles. The Vincennes' electronic equipment failed to discriminate between an Airbus and an F-14 fighter aircraft.

In both of these cases, technology forced the individual person in the loop to make a tragic error. Worse still might be a system that, because of severe time constraints, eliminates the human decision factor entirely. SDI sensor systems, for instance, must be able to identify incoming Soviet missiles and automatically trigger a defense within seconds. There is just not enough time for a human decision maker to ponder the alternatives. Bearing that in mind, consider some of the superweapons of today and tomorrow.

Superships

In many ways, Aegis is a supership and it represents modern warfare to many. Aegis epitomizes high-technology weapons systems—both the impressive capabilities and the flaws. Aegis-type ships are designed to protect an entire aircraft carrier battle group. They are able to detect threats as far as 200 miles away. Aegis electronics can track 200 targets simultaneously, arrange them according to the immediacy of the threat posed by each, and launch standard missiles to intercept the targets.

Aegis equipment includes the SLQ-32 (called "Slick 32") electronic warfare system, which is supposed to be able to recognize and identify radar emissions and it should be able to identify the Iran Airbus as a commercial aircraft. It could only do this, however, if the Airbus's weather radar or navigation radar were turned on and it may not have been. The radar screens on the Aegis do not display actual blips like conventional radar consoles, but rather, stylized symbols appear on the screen.

A less-sophisticated radar might have been able to distinguish between an Airbus and a fighter aircraft by the size of the blip. High tech was the problem rather than the solution in this case. The United States cannot step back from technology and have ship's personnel scanning the skies with binoculars or fighter pilots using their eyes to find targets like Errol Flynn in *Dawn Patrol*. Our alternative seems to be to learn our lessons and improve our technology.

Superfighters

Sometime in the mid-90s, the United States Air Force will unveil its advanced tactical fighter (ATF). It is designed to cruise great distances and fight at supersonic speeds. The proposed ATF will also be highly agile, harder to detect by radar, and able to take off and land on short runways.

Because of high fuel consumption, present-day fighters are limited to relatively slow speeds—Mach 0.85 or so for cruising and Mach 1.2 for fighting. They can maintain supersonic speeds for only minutes by using afterburners that exhaust their fuel rapidly.

The superfighter will be able to turn tighter and accelerate quicker than current fighters. ATF prototypes have already demonstrated the ability to sustain nine times their own weight during turns (9 g)—almost twice as much as today's fighters.

What about the pilot for such a vehicle? The pilot, reclining to withstand high g forces during maneuvers, will function primarily as a decision maker. Automatic systems will fly the aircraft—possibly even during combat. Such systems will keep track of the ATF position, as well as the position of adversaries. Instead of using their eyes to find targets, fighter pilots of the future will watch radar and infrared signals projected on the inside of the helmet visor. The increasing speed and range of opposition weaponry will once again be forcing the human out of the loop.

As the role of superfighter pilots becomes ever more passive, questions may be asked about their eventual replacement by robotic equipment. Fighter technology has reached the point where biology dominates. We have reached the edge of the human pilot's physical and mental ability. The reality of airwar today is speed, and relative closure speeds—two hostile aircraft approaching each other—are such that if a pilot detects a possibly hostile vehicle at a distance of 20 miles away the human element in the system has less than 30 seconds to take action.

The film *Top Gun* depicted high-performance Navy fighters in dog-fight situations, and it was great fun, but that is not the way airwars are fought. Analysis of fighter plane battles in both World Wars I and II has shown that some 80 percent of kills came not as a result of dog-fights but were of the hit-and-run type. Analysis of the Battle of Britain airwar showed that speed was more important than agility or maneuverability. Speed is life in air combat.

The proposed ATF is what is called a stand-off fighter, that is, it will be capable of combat actions beyond visual range (BVR). This means that the ATF's capability is based on sophisticated radar for identifying targets and long-range missiles for carrying out attacks. BVR air combat could require computer equipment with such a fast reaction time that the human pilot would be a liability rather than an asset.

Another area where the game is getting more automated is in underwater warfare.

Antisubmarine Warfare (ASW)

Both superpowers are well aware that a breakthrough in ASW could mean global supremacy in naval warfare. Antisubmarine warfare composes the largest slice of the U.S. Navy's budgetary pie that is allocated to a single mission. Why is ASW so important? As you saw in our discussion of the strategic forces, both the United States and the Soviet Union rely on ballistic missile-firing submarines as the ace in the hole for deterring nuclear war. But this is true only so long as the missile-carrying submarines remain virtually undetectable under the sea. The stakes in the ASW game could not be higher.

Acoustic surveillance by means of sonar (sound navigation and ranging) is the primary means of detecting, locating, and classifying submerged vehicles, so a bit about this technology is discussed next.

Sea water is opaque to virtually all types of electromagnetic radiation (except the difficult-to-transmit extremely low-frequency radio waves). Radar won't work under the ocean, and radio communication with submerged subs is not possible unless they trail an antenna on the surface, which they do not do when they are trying to escape detection.

Sound waves, however, can travel over long distances through the ocean. Sonar detection equipment uses sound waves. There are two types of sonar to consider. *Passive sonar* functions solely as a listening device. It consist of arrays of hydrophones—underwater microphones— that listen for the distinctive acoustical signature (frequency spectrum) associated with each submarine class. All submarines generate vibra- tions from the hull, the drive train, the propeller, or the pumps that are unique to that type submarine (hence the signature).

Active sonar emits pulses of acoustic energy (called pings) into the ocean and then listens for the echoes reflected from the hull or wake of a possibly hostile submarine. Active sonar provides better infor- mation than passive sonar, but it also gives away the location of the tracking sub.

The solution employed by the U.S. Navy is to use surface ships to emit the active sonar pings while a submerged attack submarine listens covertly to the echoes. Also, helicopters can drop depth charges that generate an intense burst of sound waves that bounce off a target submarine and are picked up by passive sonar arrays miles away.

In addition to attack submarines, the U.S. Navy relies on under- water hydrophone arrays to monitor Soviet submarine activity. These arrays make up what is called the SOSUS—Sound Surveillance Sys- tem—and they are located at various classified locations in the North Atlantic and Pacific Oceans to assist the U.S. Navy to detect and classify submarine movement.

Because the speed at which sound travels through water is affected by temperature, pressure, and salinity, analyzing detected submarine noises is difficult. The oceans are, it seems, noisy places, and extracting a telltale submarine noise against background noises generated by ma- rine life, shipping, and offshore drilling requires extensive computer processing.

As the technical ability to identify submarines improves, the tech- nology of quieting the submarines also improves. Advances in ASW surveillance are often offset by improvements in submarine technology, including quieter power plants, hulls made of nonmagnetic titanium, and sonar-deadening coatings. For the foreseeable future, subs and sub hunters will continue their deadly game of cat and mouse under the world's oceans.

Tank Technology

Tank and antitank technology has followed a similar course to the sub and antisub technology. First an almost impregnable tank is developed; then, almost before it can be deployed, an antitank technology is developed that makes the former pride of the ground forces obsolete.

During the 1973 Yom Kippur war, the Syrians used Soviet-built Sagger missiles to take a large toll of Israeli tanks. The "shaped charge" on the missile formed a jet of molten metal that can burn through tank armor. The answer to this threat was something called "reactive armor" that consisted of boxes of explosives hung on the outside of the tanks. When a missile hits, the armor explodes, deflecting the molten jet.

The next development was "ceramic armor," consisting of layers of ceramics that offer some protection from both missiles and shells fired by other tanks. This advantage didn't last long, however, as it was soon countered by a new type of ammunition. Instead of the long thin projectiles used in most modern guns, a new short, fat, cannonball-like projectile was developed that could shatter the ceramic.

You no doubt can see the story line now. In the United States, work has been underway on a classified project to develop still-newer armor using densely depleted uranium. Will the game end here? Not likely. The next technological advance could include "top-attack" missiles fired from helicopters or even "electromagnetic guns" designed to propel small projectiles at incredible velocities.

The important point to make in all of this is the speed of technological advancements. The U.S. M1 tank, costing $1.8 million each, was obsolete before it was ever deployed in Europe and, in fact, had to be recalled.

The blunt fact is that the tank (not just the M1 but all tanks) may be obsolete. Antitank technology is ahead of tank technology today and is likely to remain ahead (and cheaper) for the foreseeable future. It is just not possible to hide 60 tons of hot metal on the modern battlefield from the sophisticated sensors of intelligent missiles.

Consider, as one example, the range of tank guns versus the range of antitank weapons. Today's tank guns have a range of little more than a mile. Even small antitank missiles, such as TOW, are effective at ranges far in excess of that.

If you find yourself in a war in the near future, try to stay out of tanks. In fact, the lethality of modern weaponry is such that it might be better to stay off the battlefield altogether.

The Automated Battlefield

The increased deadliness of modern weapons means that people are going to be killed at a rate unimaginable in the past. Given the large number of casualties predicted for tomorrow's battlefield and the limited medical facilities available, a relatively few number of casualties can be expected to survive. A 1985 U.S. Army report recognizes this possibility and calls for the use of "corpse disintegrators" to cope with the disposal problem of so many corpses on the battlefield.

A more positive approach to the carnage of modern warfare is to remove the human from the system altogether. As modern weaponry has made the battlefield an unacceptably dangerous place to be, attention has been turning to automated warfare.

Futurologists now forecast a battlefield where opposing forces are not human but mechanized robots and automated weapons systems commanded by remotely stationed generals who follow the action on TV screens, basing their decisions on information from reconnaissance satellites, unmanned aircraft, and underground sensors. All of the action of the battle would be controlled by computers.

We may be slipping into the world of science fiction here, but the clear trend in military technology today is automation. There is no technological reason why the conventional battlefield should not become fully automated. Enemy forces can be located and identified by means of surveillance and target acquisition sensors. This information can then be transmitted to central computers for analysis, and these computers will be programmed to decide on the action to be taken. On the fully automated battlefield, the computers would select appropriate unmanned weapons and send them to their targets without human involvement. The bloodless battlefield—a war fought entirely by machines—is an idea that could catch on.

Brilliant war machines, however, are incredibly costly. The current technological arms race is an expensive game, and it is just possible that it is also a losing game. Whatever one side invents, the other side copies or invents a countermeasure for. But what is the alternative? A pullback or slowing down of the arms race—involving both conventional weapons and strategic arms—seems to be in the cards for the near future. Israeli diplomat Abba Eban once said, "Governments will always do the right thing—after they have exhausted every other option." Arms reduction may come to pass not for moral reasons, but simply because neither superpower is able to afford the luxury anymore.

START means a critical scaling down of our entire nuclear arsenal, possibly including defensive systems, but the critical element of any arms control agreement is verification. On-site inspections will be part of any future verification program, but the real burden of watching what an adversary is up to will depend on technology.

THE TECHNOLOGIES OF NUCLEAR ARMS CONTROL VERIFICATION

Verification is the major issue in arms control. Frequent reports alleging Soviet noncompliance with the provisions of various treaties and agreements reveal the need for reliable verification capabilities—either to detect Soviet transgressions or to reassure the eternally suspicious. The emphasis here is on the technological means to determine warhead numbers, nuclear test yields, and other weapons system characteristics covered by various agreements.*

The National Technical Means, or NTMs as they are called, include surveillance satellites such as the KH-11, RHYOLITE, and MAGNUM, high-tech reconnaissance aircraft, and long-range seismic detection networks.

Surveillance Satellites

Three types of satellites keep track of military activities below: low orbiters, medium orbiters, and high orbiters.

---

*Sources for this section are all unclassified and included *Deep Black: Space Espionage and National Security*, by William E. Burrows (New York: Random House, 1987).

Low Orbiters

Operating at altitudes between 70 and 350 miles, low orbiters take pictures and tap into communications signals. They can provide real-time (actual and immediate as opposed to time-delayed), close-up views of requested target areas. Resolution, that is the ability to distinguish separate objects, may be as good, if not better, than 3 inches (7.6 centimeters). They can read a license plate on a vehicle.

The powerful telescopes on these satellites are equipped with zoom lenses and computer-adjusted mirrors to compensate for the distortion caused by the atmosphere. Other sensors include infrared systems that measure radiated heat. Infrared provides effective night imagery and renders camouflage useless. Multispectral scanners employ separate lenses to shoot the same scene in different parts of the electromagnetic spectrum. It is as if the satellites take pictures of the same object through different-colored filters. Other sensors generate detailed portraits of a target's chemical makeup.

Medium Orbiters

Operating from 1000 to 10,000 miles out in space, medium orbiters function as radar ferrets and ocean surveillance platforms. They also electronically eavesdrop on some radio transmissions.

High Orbiters

These satellites travel 22,300 miles above the Earth in geosynchronous orbit (the satellite's period of rotation, 24 hours, matches the Earth's so the satellite remains in the same spot relative to the Earth). Equipped with antennas that pull in VHF, UHF, and microwave radio signals, as well as long-distance telephone traffic, sensitivity is so good that they were reported to have picked up walkie-talkie chatter between rescue workers during the Chernobyl accident.

High orbiters can also gather data on above-ground and underground nuclear tests. To accomplish this task, they are outfitted with nuclear detection systems such as particle counters, electromagnetic pulse sensors, gamma ray detectors, and X-ray measuring equipment.

High-Tech Reconnaissance Aircraft

SR-71

Until their retirement in 1989, the SR-71 Blackbirds were used to complement satellite surveillance efforts. The planes flew at altitudes approaching 19 miles (30 kilometers) and at speeds of nearly Mach 4 (four times the speed of sound and therefore literally faster than a speeding bullet—judged by Los Angeles Police Department ballistic experts to be 3500 feet per second or a little more than three times the speed of sound). Generally outfitted with frequency-specific antennas, interpretive electronic filtering equipment, and large-capacity tape recorders, the SR-71s gathered data on radio communication systems, radar units, and other electronic equipment.

RC-135

A military version of the Boeing 707, the RC-135 operates as a stand-off platform in international airspace in order to monitor Soviet major missile and space launch facilities. A battery of onboard cameras captures each liftoff and reentry with enough detail to permit accurate size, performance, and capability analysis. At the same time, advanced telemetry equipment records all signals from the reentry vehicles.

Many experts believe that it was an RC-135 that the Soviets thought they were shooting at when they downed Korean Airlines flight 007 in September 1983.

Seismic Networks

Current seismic equipment is so sensitive that it is highly unlikely that any nuclear explosion anywhere in the world could go unrecorded and unmeasured. Figure 9–3 illustrates a typical seismometer system. New seismic technology is capable of detecting blasts under one kiloton from long distances. Underground nuclear detonations display a different frequency signature than earthquakes or other movements within the Earth's mantle. By considering such variables as rock type, wave amplitude, and test site geology, computer programs can estimate weapon yields within 10 percent or better.

═══ KEY CONCEPTS ═══

▶ Deterrence and war-fighting capability are conflicting philosophies, and weapons built for one purpose do not fit the other.

▶ Restructuring our nuclear forces will require choices between competing systems such as Midgetman and MX or B-1 and Stealth. Further

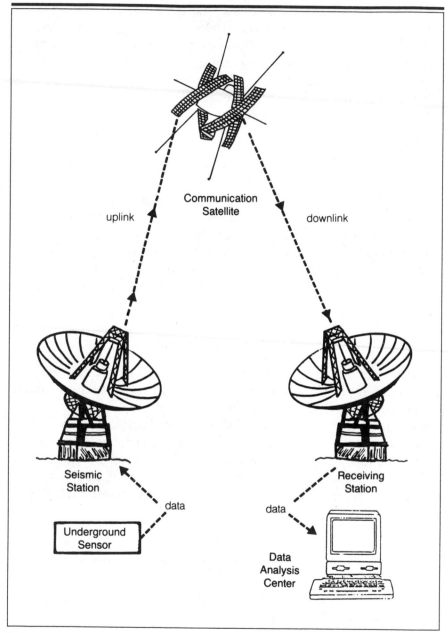

Figure 9-3 *Seismometer System Able to Monitor Underground Weapons Testing Over Long Distances*

development of cruise missiles by either side will pose a serious threat to arms control.

▶ Today's SDI is not the same space shield protecting the continental United States that Ronald Reagan proposed in 1983. Today's SDI concept is in reality a limited antiballistic missile system.

▶ Survivability of strategic weapon systems—that is, making them mobile, small, or hidden in some way—conflicts with the need for accountability in arms control agreements.

▶ High tech weaponry may be taking the human out of the loop, and that is a subject of serious concern to defense analysts.

▶ Developments in precision-guided missiles may have made many current weapons systems (such as the M-1 main battle tank, B-1B bomber, advanced fighter aircraft, and large surface ships) obsolete.

▶ Both superpowers are aware that a breakthrough in ASW technology could mean global supremacy in naval warfare.

▶ If and when nuclear arms treaties are ratified in the future, an array of reconnaissance and surveillance hardware will make cheating extremely difficult if not impossible.

APPENDIX: ANSWERS TO SELF-TEST

SPACE

1. *c* A light year is one measure of distance in space and is the distance light travels in one Earth year: equal to 5.8×10^{12} (about 6 trillion) miles. A separate and often used measure of distance in space is an astronomical unit, which is defined as the mean distance between Earth and Sun, equal to 92.81 million miles.

2. *b* A star is a celestial object that generates energy by means of nuclear fusion at its core and is therefore self-luminous. The Sun is a star. Other bodies in the universe do not generate light of their own but shine only because they reflect the star-light that falls on them.

3. *a* The approximate distance from the Sun for each planet is

Mercury	36 million miles
Venus	67 million miles
Earth	93 million miles
Mars	141 million miles
Jupiter	480 million miles
Saturn	900 million miles
Uranus	1.8 billion miles
Neptune	2.8 billion miles
Pluto	3.6 billion miles

4. *c* The Big Bang theory is a concept of cosmic history in which the universe begins in a state of high density and temperature,

both of which decrease as the universe expands. The theory is the logical consequence from the discovery that the universe is expanding.

5. *c* Black holes are regions in space into which matter is sucked by extremely intense gravitational attraction and from which nothing, not even light, can escape.

BIOTECHNOLOGY

1. *a* DNA is one of two varieties of nucleic acid (the other is RNA) found in the nuclei of cells. DNA is the chemical basis for all living things and contains a code—the genetic blueprint—describing the creature from whose cells the DNA is taken.

2. *b* Chromosomes are the threadlike substance that carries the collection of genes; the human species has 23 pairs.

3. *a* Genetic engineering is the direct manipulation of genetic material to alter the hereditary traits of a cell, organism, or species.

4. *a* In January 1989 officials from the National Institutes of Health gave scientists permission to begin gene transfer trials in connection with a cancer research project.

5. *c* The Human Genome Project will be an attempt to determine the sequence of the entire 3 billion base pairs in the human genome DNA that spell out our genetic endowment.

COMPUTER LITERACY

1. *b* A computer is a machine that manipulates the symbols of information such as numbers and letters. A computer can also be described as an electronic device designed to accept data, perform prescribed computational and logical operations at high speed, and output the results of these operations.

2. *a* A computer program is a detailed, step-by-step set of directions telling a computer exactly how to proceed to solve a specific problem or process a specific task.

3. *b* Hardware is the physical machinery of a computer system; software is the program needed to direct the operation of the machine.

4. *c* The central processing unit (CPU) is the brains of the com-
        puter. It consists of the control unit and the arithmetic and
        logic unit.
5. *b* The binary numbering system, which uses just 0s and 1s, is
        the alphabet of computers. The symbols 0 and 1 are called
        binary digits or bits. There are usually 8 bits to a byte and two
        or more bytes make up a word.
6. *d* Disks, floppy or hard, and magnetic tape are all information
        storage devices.

ENVIRONMENTAL ISSUES

1. *b* Climatologists tell us that the continued burning of fossil fuel
        and other industrial and agricultural activities increase gases in
        the atmosphere that absorb the infrared energy radiating from
        the Earth's surface, thereby trapping heat within the atmos-
        phere.
2. *b* The link between the worldwide use of chlorofluorocarbons
        and the measured losses of global ozone has been established.
        The ozone layer that encircles the Earth shields it from harmful
        ultraviolet (UV) radiation.
3. *b* Acid rain is the term used to cover precipitation of higher than
        normal acidity. Natural rain has a pH value of 5.6, and the
        lower the pH number, the more acidic the substance.
4. *d* The most effective way to control pollutants that cause acid
        rain is to burn less fossil fuel and/or reduce emissions from
        fossil fuel burning.
5. *d* Burning, burying, or dumping toxic waste have all proven to
        be ineffective for the long-term disposal of toxic waste. New
        technologies to render toxic waste harmless are being encour-
        aged by the government.
6. *c* Nevada's Yucca Mountain has been selected by the U.S. Energy
        Department as the national repository for nuclear waste ma-
        terial. A political and scientific consensus has yet to be reached
        on this issue, however.

ENERGY ISSUES

1. *c* The United States obtains by far the largest percentage of its energy (43 percent) from petroleum. About 47 percent of all the petroleum we use is imported.

2. *b* The British thermal unit or Btu is a common unit used to measure energy. It is defined as the amount of energy required to raise 1 pound of water by 1 degree Fahrenheit (specifically from 39.2 F to 40.2 F).

3. *c* Hydrogen fuel, while offering great long-range potential, faces great technical challenges, and even advocates of a hydrogen future do not foresee significant use of this fuel before 2025.

4. *a* Both fission and fusion are nuclear reactions in that they change the structure of an atomic nucleus. Fission splits atomic nuclei into smaller parts, whereas fusion joins together (or fuses) two atomic nuclei.

5. *b* Sunlight can be absorbed by rooftop collectors and used to heat water and space or it can be converted to electricity by means of photovoltaic devices.

SUPERCONDUCTIVITY

1. *d* Electricity is the flow of electrons, and the material through which the electrons flow is called a *conductor*. Different materials vary in their ability to conduct electricity. Copper is a good conductor, for instance, whereas wood and glass are poor conductors.

2. *b* The phenomenon of superconductivity occurs when a material has been chilled to an extremely cold temperature. Different materials require different degrees of refrigeration, and current research is focused on finding new materials that can superconduct at higher temperatures.

3. *b* A material is considered a superconductor when it loses all resistance to the flow of electricity. Current losses due to resistance are eliminated in a superconductor.

4. *a* A major breakthrough in superconductivity occurred when a material was discovered that would offer no resistance to the flow of electrons at temperatures above 77 Kelvin (-320.8

degrees Fahrenheit). At temperatures below 77 Kelvin, relatively expensive helium must be used as a refrigerant, whereas at temperatures above 77 degrees Kelvin much cheaper liquid nitrogen may be used.

5.   *d*   While many technical problems remain, superconductivity has great potential for improving the efficiency of power generation and distribution; advanced electronic systems, including superfast computers; and transportation systems, including levitating trains.

High-Technology Medicine

1.   *a*   A CAT scanner produces a cross-sectional view of tissues within the human body as compared to conventional X rays, which view the body from only one angle.

2.   *d*   Sonography uses high-frequency sound waves to look within the body. Under specific circumstances, sonography can obviate the need for diagnostic techniques requiring radiation of some type.

3.   *b*   Digital subtraction angiography or DSA is an imaging technique that produces clear views of flowing blood or its blockage by narrowed vessels. DSA requires only one-third the amount of dye necessary in a conventional angiogram and is currently the technique of choice in revealing narrowing or other problems in blood vessels.

4.   *c*   Positron Emission Tomography (PET) is an imaging device that differs from CAT in that PET scans provide functional perspectives of biochemistry occurring in living tissues.

5.   *b*   Artificial joints (hip, shoulder, elbow, knee, and ankle) can now be designed to order and made from such materials as stainless steel, plastic, and chromium.

6.   *a*   Medical lasers are widely used in medicine today. They can be used for some types of surgery, or to halt internal bleeding, or to vaporize some kinds of abnormal growths or tumors. Lasers are not an imaging device, however, and they cannot be used in place of X rays.

TRANSPORTATION TECHNOLOGY

1. *b*   The First and Second Laws of Thermodynamics govern the ability of an automobile engine to convert chemical energy in the fuel to the mechanical energy that propels the car. They thus set the theoretical limits of engine efficiency.

2. *b*   Air resists the movement of a vehicle passing through it. The resisting or "drag" forces increase with the square of the vehicle speed: twice the speed produces four times the resistance.

3. *b*   A Mach number, named after Austrian physicist Ernst Mach, measures the ratio of an object's speed to the speed of sound (740 miles per hour at sea level). Mach 2 would be equal to 1480 miles per hour.

4. *b*   Antilock braking systems (ABS) are more efficient and safer than conventional systems because a vehicle will stop faster and remain steerable when the brakes are applied many times per second, instead of locking up tight.

5. *b*   Studies of the actual operation of air bags reported no cases where air bags failed to deploy in a crash or deployed inadvertently. This reliability rate is far higher than that of such safety features as brakes, tires, steering, and lights, which have shown a failure rate of up to 10 percent.

6. *b*   In a fly-by-wire system, the pilot communicates (by wire, hence the name) commands to a computer, which then activates the hydraulic cable systems needed to control the aircraft. The main advantage to the fly-by-wire control system is that the electrical control network is ideally suited to the more extensive use of computers. Planes under computer control can respond more quickly to turbulence and other changes in flying conditions.

SUPERWEAPONS AND ARMS CONTROL

1. *c*   Mutually Assured Destruction (MAD), sometimes called the "balance of terror," is a technological stalemate in which neither superpower can attack the other without incurring unacceptable damage to itself. Critics refer to this concept as MAA (mutually assured anxiety).

2. *b*   U.S. strategic forces are divided into what is called the triad: air-based, sea-based, and land-based systems. In order to ensure a reliable retaliatory force, it was not considered wise to place all our reliance on any one type of nuclear weapons delivery system.

3. *b*   The original SDI or Star Wars concept was described as a shield that could protect the United States "just as a roof protects a family from rain." It was proposed as a defense against nuclear attack so perfect that it would render nuclear weapons impotent and obsolete.

4. *d*   The increased lethality of modern warfare means that people are going to be killed at a rate unimaginable in the past. Automated warfare may make it possible in the near future to remove the human element from the battlefield in response to this increased deadliness.

5. *c*   Sonar equipment uses sound waves that can travel over long distances through the ocean. Passive sonar functions solely as a listening device, whereas active sonar sends out a pulse of sound and then listens for the returning echo.

6. *c*   New seismic technology is capable of detecting from long distances blasts under 1 kiloton and differentiating nuclear explosions from earthquakes or other movements of the Earth's mantle.

INDEX